BATTLEFIELD
THE WEAPONS OF MODERN LAND WARFARE

Consultant editors
Christopher Foss and Ian Hogg

ORBIS · LONDON

Acknowledgments

Photographs were supplied by:
AB Bofors, Aerospace, Alvis, Antenne National, A.P.N., Australian Army, British Aerospace, Bundeswehr, Cadillac Gage, Camera Press, C.O.I., Creusot-Loire, Crown Copyright, C.T.K., DAF, E.C.P.A., Educational and TV Films, Engesa, Eshel-Dramit, Federal German Army, Ford, Christopher Foss, General Dynamics, GIAT, GKN, Heeresfilm, John Hillelson Agency, Ian Hogg, Hughes Aircraft Corporation, Robert Hunt Library, Imperial War Museum, Italian Army, Kraus-Maffei, Krupp Mak, Martin Marietta, M.B.B., M.O.D., Mowag, National Archives, John Norris, Novosti, Popperfoto, Quittard, The Research House, Rex Features, Rheinstahl, Rockwell International, Short Brothers, Soltam Ltd, Frank Spooner Pictures, Swissair, Tass, Thomson-C.S.F., Thyssen Henschel, Tonkin, UK LF, U.P.I./Bettmann, US Army.

Editor Richard Williams
Designer Michael Moule

First published in Great Britain 1986 by Orbis Book Publishing Corporation
A BPCC plc company

Printed in Italy

ISBN: 0–85613–846–0

Title page: British 105mm Model 56 pack howitzers in action.

CONTENTS

MODERN ARMOURED FIGHTING VEHICLES

Although tanks were first used successfully during World War I it was not until the 1920s and 1930s that armies gave serious thought to mechanisation. It was the Germans who first took the message to heart and developed the necessary vehicles, weapons and tactics. German mechanised units supported by tactical aircraft were responsible for the successful German conquest of Europe in 1939-40. World War II proved that the tank had a major role to play on the battlefield, but that it had to be supported by other arms such as infantry, artillery and engineers and these had to work in close cooperation with ground-attack aircraft.

Since World War II there have been armed conflicts in almost every part of the world with the exception of Europe. In the Korean War both sides used tanks to a varying degree, but the nature of the terrain made it difficult to use them as part of a combined arms team, and they were usually relegated to the role of fire support. The North Koreans and Chinese used the Soviet T34/85 tank and SU-100 and SU-76 self-propelled guns, while United Nations forces in the South had at their disposal Shermans, Chaffees, M26 Pershings, M46s and the British Centurion.

The Israeli Army has had more successful battle experience in mechanised warfare since World War II than any other country. In both the 1956 and 1967 Middle East conflicts, Israeli armoured units played a key role and the latter campaign was over in just six days with very few Israeli casualties. The 1973 War was much tougher as Israel was attacked on two fronts at once by Egypt and Syria. In the early days of this conflict the Israelis suffered a number of defeats both in the air and on the ground, but they eventually fought the Egyptian and Syrian armies to a standstill before forcing them back. Many of the initial Israeli defeats were due to using tank units without infantry, artillery and engineer support, in which they fell easy prey to the Egyptian infantrymen armed with the Soviet-supplied RPG-7 and Sagger anti-tank guided weapons. Experience in the Middle East has led to Israel developing its own tank, the Merkava.

The French made considerable use of US-supplied armoured vehicles in Vietnam but the South Vietnamese and Americans took some time to appreciate the value of armour in the war. Initial trials with American M113 full-tracked armoured personnel carriers were such a success, however, that they were soon being employed on a large scale on search-and-destroy missions. Many of these vehicles were specially modified for use in Vietnam and fitted with additional armament. The South Vietnamese used the Chaffee light tank in the fire-support role and this was soon supplemented by the M41, while the United States used both the M48 tank and the M551 Sheridan light tank. The latter was not a success in South Vietnam as there were problems with its armament, and its armour proved to be too thin to provide much protection against mines. Of all the weapons employed against armoured vehicles in South Vietnam it was the mine that caused the greatest casualties and by the end of the conflict no easy solution to this problem had been found. The Americans also deployed self-propelled guns, tank destroyers and self-propelled anti-aircraft guns in South Vietnam, using the latter for fire support. On the other side the North Vietnamese also used tanks, especially the PT-76 light amphibious tank, and a number of tank battles took place during the final advance to Saigon.

There have been a number of conflicts between India and Pakistan since the two countries came into existence after World War II. In the 1960s Pakistan was supplied by China and the United States while India received its equipment from Britain and

Above: A British Chieftain MBT advances at speed during a NATO training exercise. Great things were expected of the Chieftain when it entered British Army service in the 1960s but the tank was plagued with engine problems; the L60 650bhp power pack proved far too small for an AFV of the Chieftain's size. This weakness was eventually rectified and later models were equipped with 1200bhp Rolls-Royce engines.

France, and Indian Centurions had the edge in engagements with Pakistani US-supplied M47/M48 tanks. In the last 20 years India has made determined efforts to build up its arms industry and from 1965 produced the Vijayanta MBT (main battle tank) designed to meet the requirements of the Indian Army by the UK-based firm of Vickers. More recently India has turned to the Soviet Union and is now making the T72 under licence.

Armoured vehicles of one type or another have been used in many of the more recent conflicts. The British used the Scorpion/Scimitar vehicle in the recent Falklands War but these were employed in the fire-support role and never engaged the Argentinian Panhard AML-90 armoured cars deployed around Port Stanley. South Africa has used its Ratel (6 × 6) infantry fighting vehicles and Eland (4 × 4) light armoured cars in its many deep-penetration raids into Angola and these have often encountered Soviet-supplied vehicles such as the old T34/85 tank and BRDM-2 armoured cars. Armoured vehicles of one type or another have been most widely used for internal security operations, for example in South America and Northern Ireland. In most cases these are wheeled vehicles as they are cheaper to maintain and operate; they are also more acceptable than tanks from a political point of view.

Apart from the unique Swedish S-tank with its fixed 105mm gun and the Israeli Merkava MBT, most tanks today have an almost identical layout with the driver at the front, turret in the centre and the engine and transmission at the rear. The Soviet T64/T72 tank has an automatic loader which has enabled the crew to be reduced from the normal four to just three: commander, gunner and driver. Future Western tanks will probably be much more compact than in the past, making them lighter and more difficult to hit.

In recent years there have been significant advances in tank technology. Guns now fire ammunition with greater velocity and higher penetration characteristics, and tanks have more powerful and fuel-efficient engines, improved suspensions, and new types of armour protection such as the British Chobham armour and the Israeli reactive armour. There have also been significant advances in fire-control systems with the introduction of computers, laser rangefinders and day/night sights.

Armoured cars have been around since well before the introduction of the tank during World War I. Armoured cars and light tanks have traditionally been the eyes and ears of the brigade, division or

Above: The business end of a Chieftain, the 120mm L11 gun. Initially this gun was aimed using a 12.7mm ranging machine gun but this was superseded by the vastly more sophisticated Barr and Stroud laser rangefinder, an improvement in accuracy which allowed targets to be engaged at the gun's maximum range of 4000m (4400yd). Additional armament comprises two 7.62mm machine guns.

corps commander, as in a fluid battle they find out the strength, position and future intentions of the enemy so that the commander can deploy his forces to meet the threat. In many armies this is still the role of the armoured car or light tank, but in some more sophisticated armies this role is being taken over by other systems such as helicopters or RPVs (remotely piloted vehicles). An RPV, mounted with a camera, can provide the commander with a detailed picture of what is happening on the battlefield as the action is taking place.

While the tanks may take the ground it is the infantry who have to hold onto it so most armies now have tracked or wheeled infantry vehicles, or a mixture of both. Generally tracked infantry vehicles have better cross-country performance and are better armed and armoured, while the wheeled vehicles are cheaper to purchase, maintain and operate and have greater strategic mobility as their wheels give them a higher road speed than their tracked counterparts. Very often the APC (armoured personnel carrier) is the basic member of a complete family of vehicles that also includes command-post vehicles, mortar carriers, anti-tank vehicles, specialised fire-control vehicles, repair and recovery vehicles, and so on.

In addition to tanks, armoured cars and APCs, Nato armies have had self-propelled guns for a long time and these are being issued on an increasing scale to the armies of the Warsaw Pact. The French 155mm GCT has an automatic loading system that enables a high rate of fire to be achieved. With the introduction of sophisticated detection equipment this becomes very important, as self-propelled artillery systems will only have a limited firing time before they have to move off to a new position. Self-propelled anti-aircraft systems are deployed by many armies to protect their forward elements from attack both by aircraft and helicopter. The engineers also employ a host of specialised vehicles to support their armour, including recovery vehicles, bridgelayers and mine-clearance devices.

Armoured warfare is in a continual state of flux with the introduction of new equipment and new tactics. For some years a few people have believed that the tank has no future. When employed in isolation this is probably true, yet when used in conjunction with all the other elements of a battlefield force, it can still be the key to victory.

Above: A column of Warsaw Pact T62s pauses during manoeuvres. The T62 is armed with a 115mm main gun plus two machine guns.

Below: An M107 self-propelled artillery piece of the British Army. This SPG carries a 175mm gun with a maximum range of 32,700m (35,750yd).

ABRAMS M1

By late 1984 over 2000 Abrams M1 main battle tanks (MBTs) had been produced by the Lima Army Tank Plant and the Detroit Arsenal Tank Plant, both of which are operated by the Land Systems Division of General Dynamics, who took over Chrysler Defense several years ago. The M1 has been evaluated by Switzerland and Saudi Arabia but the only user at present is the US Army. The first M1A1 production vehicles were scheduled to be completed in 1985; these have a number of major improvements over the original M1, the most significant being the replacement of the original 105mm rifled gun by the West German-designed 120mm smooth-bore gun.

The Abrams M1 MBT (main battle tank) came into operation in 1980 and represents the latest thinking in American tank design. The M1 was named after General Creighton W. Abrams, the former Commander-in-Chief in Vietnam, and Chief-of-Staff, who approved the project personally in September 1972.

The United States had been a partner with Germany in the MBT-70 project, a joint-development effort begun in August 1963 aimed at helping standardise Nato's advanced tank requirements. Conceived and nourished by experience gained after World War II both partners were to invest jointly in designing what was intended to be the most advanced armoured fighting vehicle in the world. But although prototypes were ready for combat evaluation in 1967 in the shape of a sleek 50-tonne, three-man tank, problems soon arose.

While the Americans required the new tank for worldwide deployment, the Federal Republic of Germany favoured a design suited to warfare in central Europe. Differing firepower concepts were also a prominent issue. The Germans, from their World War II combat experience, favoured a high-velocity long-range large-calibre gun, while the Americans wanted to fit the Shillelagh 152mm gun/launcher system capable of firing a variety of ammunition types as well a Shillelagh missile.

The differences between the two nations seemed

Previous page: The M1 crashing through rough country at speed.
Above: The MBT-70 prototype – a highly unusual design with a hydropneumatic suspension system and a three-man crew, with the driver's position on the left side of the turret in a contra-rotating ring.
Below: The M1 as it finally appeared, with a 105mm main armament and a four-man crew.

irreconcilable and in 1970 the US Congress ordered the joint project to be abandoned and allocated special funds for a new national project. As a result, each country went its own way, the Germans ending up with their powerful 120mm-gun Leopard 2, and the Americans going along the painful road which finally produced the M1.

The Americans began work on an 'austere' version of the MBT-70, to be known as the XM803, but this too was found to be unsatisfactory and the project was cancelled only a year after its inauguration. An urgent solution had to be found if the United States was not to be left far behind in modern tank warfare – and time was running out fast, as news of the new Soviet tank designs filtered through Allied intelligence channels. In February 1972 the US Army set up a special task force, which included user, trainer and development agencies aimed at cooperating closely in the development of the new MBT.

Following the presentation of proposals in 1972, contracts were awarded in 1973 to the Chrysler Corporation and General Motors, who both set about developing prototypes to be designated XM1. In 1976 both prime contractors presented their pilot vehicles for initial testing.

In November 1976 Chrysler was selected as the winner of the validation phase, and a full-scale engineering contract was awarded, worth $196 million. An interesting feature of the Chrysler power-plant was that it was an Avco-Lycoming gas-turbine engine, the first of its kind to power a battle tank. The turbine has better acceleration and power than conventional engines and is very economical from a field maintenance point of view, although present models consume more fuel than the high-powered diesel engines of other tanks.

In February 1980 the US Army accepted delivery of the first two production tanks in a ceremony held at

Above: An M1 with its gun traversed to the left. Below left: The pilot model of the XM1 prototype which was revealed to the world in 1978. Below right: The M1 showing its teeth, with the driver's position at the centre of the tank clearly visible.

the army production line at Lima, Ohio. Designated the M1, the US Army has ordered over 7000 of these tanks and they will provide the advanced edge of America's armoured might for the next two decades at least.

An impressive armoured vehicle, the M1 is also the most expensive tank ever produced, with its unit price surpassing $1.55 million. With a top speed of 72 km/h (45mph) on roads and up to 50 km/h (31mph) cross-country, the M1 is a fast-moving tank and a marked improvement over the existing M60A3. The same goes also for the silhouette, which is extremely low, enabling the tank to find better cover from hull down positions. The M1's ability to survive battlefield conditions has been enhanced by its quick acceleration – up to 32 km/h (20mph) in six seconds – which allows the tank to make sudden spurts from one piece of cover to the next. Also useful in this respect is the M1's comparatively low noise and smoke signature.

As in the German Leopard 2 and the British Challenger, the M1 is protected by compound armour technology. This covers the front hull and turret structure, balancing the rest of the tank with varying thicknesses of regular armour, ranging from 3.25mm (⅛in) to 125mm (5in). Spaced armour plating is added in vulnerable areas to decrease the effect from chemical action rounds such as HEAT and HESH.

The first production batch mounts the US 105mm M68 tank gun, which is a variant of the British 105mm L7 series manufactured under licence in the United States. Already fitted to the M60 tank it is capable of firing most ammunition types. However, it is planned to upgun the tank with a more powerful main armament in the mid-1980s. This will be the 120mm Rheinmetall smooth-bore gun as fitted to the German Leopard 2.

The tank's fire-control system is very advanced: gun and turret drive are hydraulically powered with the gun's elevation and depression achieved by a hydraulic cylinder; turret traverse is powered by a hydraulic motor–gearbox combination. The M1 features a Hughes laser rangefinder and a back-up solid-state digital computer, with an interior self-checking fault control. This allows the gunner to select the target, fix it with the rangefinder and then press his

Above: The driving compartment with T-bar clearly visible. Below: The interior, showing the breech of the 105mm gun.

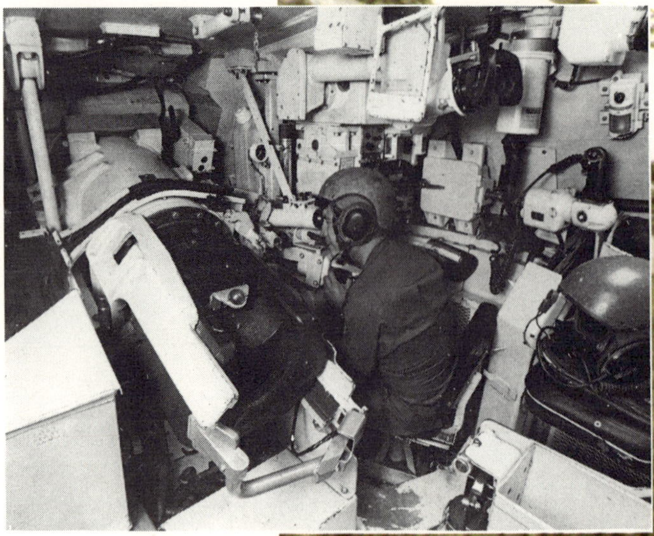

Left: A side-view of the tank with anti-shaped-charge plates protecting the suspension. Below: An XM1 ploughs across a field during its proving trials. The powerful 1500 horsepower gas-turbine engine and rugged torsion-bar suspension will provide production M1s with the ability to cross broken terrain at high speeds – a great advantage for survival on the battlefield.

fire switch. The computer then takes over and makes the necessary calculations and adjustments to ensure a hit.

A 7.62mm machine gun is mounted coaxially with the main armament and a second is fitted at the gun-loader's position on the turret. The commander has a 12.7mm anti-aircraft machine gun also fitted on the turret.

Crew protection against hits is improved by greater obliquity of the hull and turret surfaces and by armoured skirts over the suspension. Several protective innovations will improve the chances of survival following eventual penetration. Ammunition stowage has been compartmentalised with 44 main-gun rounds carried in the bustle behind sliding armour doors. Eight main-gun rounds are stowed in a compartment in the hull and three more on the turret floor, protected by spall plates. In the event of a hit, the blast of the explosion is vented upwards and out of the bustle by specially constructed blowout plates, thus directing the detonating rounds away from the crew compartment. During operational tests at the Aberdeen Proving Ground, a prototype XM1 was fully loaded with fuel and ammunition and subjected to various types of fire. Not only was it not destroyed – it was later driven away under its own power.

Although the shape of the turret is aimed at achieving a low silhouette over the hull, allowances for the crew's comfort and efficiency have prevented it from reaching the close-fitting dome shape of Soviet tanks. Moreover, there are resulting high-explosive pockets in front and at the sides, especially when the turret is traversed. The ability of the tank crew to work efficiently within the confines of an armoured vehicle under difficult conditions is of the greatest importance on the battlefield. The problem is to ensure that the tank's combat efficiency is not seriously impaired, and that costs do not become prohibitively expensive. The Soviet-built tanks of the Warsaw Pact are reliable and tough battlefield vehicles – as well as being relatively easy to produce – but the strains involved in operating them are considerable and would almost certainly have adverse consequences during sustained combat.

The driver's position is situated at the front centre of the tank and is operated from a semi-reclining position when driving with the hatch closed. Steering is done by a rotating motorcycle-type T-bar which actuates the steering lever, with grips for throttle and fuel management. A centre periscope and image intensification periscope for night driving is provided.

Above: The XM1 prototype painted with the 1976 camouflage colour scheme in preparation for combat trials.

One of the most crucial questions arising on the present validity of the M1 is the effectiveness of its gun. While the fire-control system and advanced optical equipment may well ensure a high rate of first-hits at medium ranges, it remains debatable whether the present 105mm gun, even with its advanced ammunition, would cope with the frontal armour of the Soviet T64 and T72 tanks. However, the adoption of a 120mm smooth-bore gun should improve the battlefield effectiveness of the M1.

Two more criticisms that have been levelled at the M1 are poor engine reliability and high fuel consumption. Exhaustive field testing seems to have proved the reliability of the Avco-Lycoming turbine, although information is far from complete. Tactical advantages outweigh the increased fuel consumption, especially in close or medium-range battle situations that would be likely to develop in central Europe, where movement is limited and staying power is emphasised. In fact, although the M1 was developed to meet American worldwide requirements, it will prove a highly effective weapon on the European battlefield should war break out between the Warsaw Pact and Nato. Alongside the German Leopard 2 and the British Chieftain and Challenger tanks the M1 would become a central element in Nato's ability to block the Soviet tank assault and, of course, would form the spearhead of any Nato armoured counter thrusts.

The M1 has yet to prove itself in combat, but given its sophisticated electronic equipment and good armour protection, when armed with a 120mm main gun it should be a match for any main battle tank.

Abrams M1 Main Battle Tank

Crew 4
Dimensions Length (gun included) 9.77m (32ft); width 3.65m (11ft 11in); height 2.89 metres (9ft 1in)
Weight Combat loaded 54,432kgs (119,050lbs)
Engine Avco-Lycoming AGT-T 1500 HPC gas turbine developing 1500hp at 3000rpm

Performance Maximum road speed 72km/h (45mph); maximum cross-country speed 50km/h (31mph); range (road) 490km (304 miles); vertical obstacle 1.24m (4ft); trench 2.74m (9ft); gradient 60 per cent; fording 1.22m (4ft), with kit 2.36m (7ft 9in)

Armour Chobham compound type and spaced; details classified
Armament One 105mm M68 gun (55 rounds); one 7.62mm machine gun co-axial with main armament; one 12.7mm machine gun on commander's cupola; one 7.62mm machine gun mounted by loader's hatch; six smoke dischargers on either side of turret as well as integral smoke generators

A first in tank engine development, the Avco-Lycoming gas turbine has aroused considerable interest in tank-design circles. If the engine proves a success for the Americans then the Israeli IDF may well install the turbine in the Merkava MBT when it is due for upgrading. Although residual doubts still remain regarding its battlefield reliability, its advantages are many. Ease of maintenance is a particular feature and it has been claimed that the complete engine can be replaced in less than half-an-hour. A great advantage of the engine is that it allows a variety of fuels to be used, including petrol, diesel and even jet fuel. During confused campaign operations such flexibility would be highly beneficial, and – albeit to a limited degree – would help solve the logistical nightmare that bedevils Nato's heterogeneous organisational structure.

CENTURION

The Centurion is generally accepted to be the best tank that Britain has ever produced. Design work started in 1943 and in 1984 the tank was still in front-line service with Denmark, Israel, Jordan, the Netherlands, Kuwait, Singapore, South Africa, Somalia, Sweden, Switzerland and the United Kingdom (not in its original tank role). In many of these countries the Centurion has been modernised with a new fire-control system, 105mm gun, new engine and transmission. These modifications have given the Centurion a new lease of life, and, firing the new APFSDS (Armour Piercing Fin Stabilised Discarding Sabot) 105mm round, it can take on the latest tanks.

In over three decades of active service the British Centurion main battle tank (MBT) has proved itself one of the finest and most durable armoured fighting vehicles produced since 1945. Combining reliability with a powerful main armament and effective armour protection, the Centurion has been engaged in wars fought throughout the globe – in Korea, India, the Middle East and Vietnam – and has gained the respect of both friend and foe alike. The great strength of the Centurion – like many other successful weapons and vehicles – has been its ability to be upgraded to incorporate new developments in weapon technology and thus remain one step ahead of obsolescence. The developmental range of the Centurion was demonstrated by the 25 separate variants that were produced in a production run of over 4000 tanks that spanned the period from the end of World War II to 1961.

Plans for the Centurion first took shape in 1943 when the existing British cruiser tanks, the Cromwell and its successor the Comet, had reached the limit of their design potential. Mounting 75mm and 77mm guns respectively, they were inferior in terms of armament and armour to the more powerful German tanks then entering service. To counter this threat, design work began in mid-1943 on the Heavy Cruiser A41, later to be known as Centurion. The British Army considered the following characteristics to be of the greatest importance: reliability; gun and armour configuration combat ability; cross-country performance; ease of maintenance.

The earlier British cruisers had suffered a depressing record of mechanical unreliability. Many had been lost in the Western Desert due to trivial breakdowns that strained recovery facilities. The adoption of the Meteor engine and the Merritt-Brown gearbox in the Cromwell and Comet overcame many of these earlier deficiencies; consequently, this successful combination was installed in the A41.

Though consideration had been given to the 77mm gun because of its superior high-explosive capability, a main armament capable of defeating the German Tiger and Panther tanks was deemed essential, and so the 17-pounder, which had proved its worth in anti-tank roles, was adopted. Various combinations of auxiliary weapons were considered, including a quick firing 20mm Polsten gun or two 7.92mm Besas in a ball mounting alongside the main gun, together with a machine gun in the turret rear. The 20mm Polsten, used in the first prototypes, was replaced by a single 7.92mm Besa in later prototype vehicles. The weight of armour was determined by the need to withstand the formidable 88mm anti-tank gun employed by the Germans in heavy tanks and in ground roles. In view of the increased protection afforded by a sloping glacis plate, the hull machine gunner, characteristic of earlier British designs, was dispensed with and his space used for ammunition stowage.

Performance was regarded as secondary to both reliability and effective combat characteristics. Good cross-country ability, comparable to that of the Comet, was, however, deemed necessary – even at the expense of high road speed. Experience having shown the necessity of a high reverse gear, the gearbox was modified accordingly. The hull, built with sloping sides to minimise the effects of mine damage, as well as the vehicle's increased weight, precluded the use of the Christie-type suspension of the earlier cruisers; a modified Horstman type was therefore adopted. Skirting plates were added to protect the suspension against the effects of hollow charge anti-tank weapons, common during the closing stages of the war.

One of the great virtues of the American-built Sherman tank was its ease of maintenance under field conditions; this meant that it could be repaired and

Previous page: Centurions of the Israeli Defence Forces in the Sinai Desert. Top: A Centurion Mark 5 of the 14/20th Hussars. Based on the Mark 3 design, the Mark 5 was armed with a 20pdr main gun. Above: A Centurion Mark 3 of the 8th Hussars lurks in a defensive position overlooking Seoul on the Han River. The Korean War marked the combat debut of the Centurion, where it gained a reputation as a reliable tank with a gun of deadly accuracy.

Top: The first prototype of the A41; developed early in 1945 this tank was to evolve into the Centurion. Alongside the 17pdr main armament is the 20mm Polsten gun, though this secondary armament was not used in subsequent prototypes and was replaced by a co-axial machine gun.

returned to the battlefield with a minimum of delay. Thus, attempts were made to ensure the accessibility of major components within the Centurion. An interesting innovation was the provision of an auxiliary charging engine to give reserve electrical power for the radios and gun-control equipment.

In May 1944 the mock-up of A41 was viewed at AEC Ltd and an order placed for 20 prototype models. Extensive running trials of test rigs began in September and by the following January production of the prototypes was under way. The first six prototypes, sent to Germany in May 1945, were intended to go into action with the Guards Armoured Division, but the war ended before this could be achieved.

Once the basic design was set, a limited contract was placed in 1946 for 100 vehicles designated Centurion Mark 1. Development continued, however, resulting in a number of modifications which could not be readily embodied in the production of the Mark 1. Under the design parentage of Vickers Armstrong, the Centurion Mark 2 was introduced in the summer of 1946. The new model incorporated a fully cast turret, replacing the earlier fabricated one, and a coaxial machine gun integral with the main armament as well as stabilised gun-control equipment in both elevation and azimuth. A number of other important features included a commander's vision cupola, increased hull armour and a periscopic gunner's sight in the turret roof.

However, the army did not consider that the Centurion could satisfy the current requirements for a 'universal' tank suitable for both independent armoured operations and infantry support. Such a concept, proposed by Field-Marshal Montgomery, led to the development of the A45, intended to supersede the Centurion. As the specifications called for a tank capable of a number of roles such as bridge-laying, dozing, mine flailing and swimming, it was thought that the Centurion could not fulfil this requirement without extensive redesign. The prototype – now designated FV201 – which appeared in 1948 was based on a more heavily armoured version of the Centurion, with which it shared a number of common features, including the turret and main armament.

As it happened, it was decided not to abandon the Centurion. Accordingly, in 1948, the Mark 3 – mounting a new 20-pounder gun – was introduced. This version, similar to the Mark 2, featured a hull shortened by 10cm (4½in) and redesigned transmission covers. Other improvements, including more advanced gun-control equipment and minor engine modifications, increased power output and improved reliability.

The Centurion Mark 4 was to have been a close-support version mounting a 95mm howitzer as its main armament. But in early 1949 – before production could begin – it was realised that the Mark 3 could fulfil the role with the introduction of the new 20-pounder high explosive and smoke ammunition. The

Mark 4 was then cancelled.

By 1951, production of the Mark 3 had reached the rate of 20 a month. During this period approximately 250 modifications were approved, consistent with experience gained from conditions in the field. These, though of a minor nature in themselves, combined to achieve considerable overall improvement. They included an additional guide roller, removal of the two-inch bombthrower, relocation of the loader's hatches and episcope, and elimination of the turret rear escape hatch. Earlier vehicles were constantly updated to current standards including the Mark 2s, which were upgunned with the 20-pounder to become Mark 3s.

The most pressing requirement, however, was for an increase in the Centurion's radius of action. This had dropped to barely 100km (60 miles) as the vehicle's weight rose from the A41's original 46 tonnes to the 51 tonnes of the Mark 3. To remedy this shortcoming, jettisonable fuel drums, similar to those carried by Soviet tanks, were fitted to the rear of the hull. An armoured mono-wheel trailer with 200 gallons of fuel, developed in 1952, proved unpopular with tank crews, as the increased length made manoeuvring difficult. A satisfactory compromise was found in a 100-gallon armoured fuel tank bolted to the hull rear; this was subsequently fitted to many early Centurions.

In the interests of Nato standardisation, the coaxially mounted Besa machine gun was replaced with the .3in Browning; vehicles so fitted were designated Centurion Mark 5s. Apart from an adaptation to the gun mounting, they were identical to the Mark 3s, which were retrospectively modified in 1955.

The Centurion made its combat debut during the Korean War and the first armoured engagement took place in February 1951 when a Centurion Mark 3, supporting an American patrol along the Han River, knocked out a Chinese tank at a range of 2750m (3000yds) with its second shot. In the rugged mountainous terrain of Korea the good hill-climbing performance of the Centurion was a decided asset; it could climb seemingly impossible hills to bring direct fire to bear against the enemy – an ability much envied by the crews of the US Patton and Sherman tanks.

As the conflict bogged down into positional warfare, the Centurion assumed an important infantry-support role. As the Centurion's direct fire was more accurate than that of conventional artillery, valuable support could be given to infantry attacks against the enemy's hillside positions. The accuracy of the Centurion's 20-pounder gun allowed long-range destruction of enemy bunkers and observation posts, a process known as 'posting letters'. Centurion tank crews claimed to be able to put a high explosive round through an opening two feet by one at a range in excess of 1800m (2000yds).

When the Korean War ended, overseas sales orders increased remarkably: the Centurion has seen service with the armies of Australia, Canada, Denmark, Egypt, Holland, India, Iraq, Israel, Jordan, Kuwait, New Zealand, South Africa, Sweden and Switzerland.

In January 1952, an extensive redesign programme was authorised under the auspices of Leyland Motors. The resulting Mark 7 embodied a new hull with increased fuel capacity which, sited internally within the tank, effectively doubled its range. Other principal features were lighter cover louvres for the engine and transmission, improved ammunition stowage with more 'ready' rounds and a loading port in the hull, an integral turret floor, unified screw threads, revised drive controls and larger headlights. A mock-up of the new vehicle was inspected in November 1952; production began in late 1953.

The Centurion Mark 8, introduced in September 1955, featured a revised turret design incorporating a resilient gun mounting to lessen the likelihood of the trunnion pins shearing under impact from armour-piercing projectiles. A new fire-control system, with elevation by chain drive instead of gears, was fitted, as well as an automatic sensor which engaged the stabilisation system when the tank exceeded a speed of 3.2km/h (2mph). Better protection was provided for the commander by means of a new cupola with a split hatch which could be elevated in an umbrella position to give overhead cover while retaining the advantages of direct vision. The cupola also enabled

Top: Supported by armoured personnel carriers a Centurion Mark 13 pushes forward on a training manoeuvre. The Mark 13, like other later Marks of the Centurion, was armed with the L7 series gun, a highly successful design that has been used by the Americans in their M48 and M60 tanks and by the Israelis in the Merkava MBT.

Centurion Tank Series Main Variants

Mark	Main Armament	Remarks
A41	17pdr (76.2mm)	
Mark 1 (A41*)	17pdr	
Mark 2 (A41A)	17pdr	
Mark 3	20pdr (83.4mm)	Saw service in Korea; Most later modified to Mk5 standards.
Mark 4	95mm howitzer	Proposed model; later cancelled
Mark 5	20pdr	Based on Mk3 but with a new machine gun mounted on commander's cupola plus modified gun and fire-control equipment
Mark 5/1 (FV4011)	20pdr	Increased glacis-plate armour
Mark 5/2	105mm	Upgunned Mk5
Mark 6	105mm	Upgunned and uparmoured Mk5, some models with an extra armoured fuel tank on hull rear plate
Mark 6/1	105mm	Mk6 with infra-red driving and fighting equipment; stowage basket on turret rear
Mark 6/2	105mm	Mk 6 with .5in ranging machine gun
Mark 7 (FV4007)	20pdr	Increased fuel and ammunition stowage
Mark 7/1 (FV4012)	20pdr	Uparmoured Mk7
Mark 7/2	105mm	Upgunned Mk7
Mark 8	20pdr	Resilient mantlet mounting for main gun (fitted with fume extractor on barrel); new commander's cupola and fire-control equipment
Mark 8/1	20pdr	Uparmoured Mk8
Mark 8/2	105mm	Upgunned Mk8
Mark 9 (FV4015)	105mm	Upgunned and uparmoured Mk7
Mark 9/1	105mm	Mk9 with infra-red driving and fighting equipment; stowage basket on turret rear
Mark 9/2	105mm	Mk9 with .5in ranging machine gun
Mark 10 (FV4017)	105mm	Upgunned and uparmoured Mk8 with automatic stabilising system
Mark 10/1	105mm	Mk10 with infra-red driving and fighting equipment; stowage basket on turret rear
Mark 10/2	105mm	Mk10 with .5in ranging machine gun
Mark 11	105mm	Mk6 with .5in ranging machine gun; infra-red driving and fighting equipment; stowage basket on turret rear
Mark 12	105mm	Mk9 with .5in ranging machine gun; infra-red driving and fighting equipment; stowage basket on turret rear
Mark 13	105mm	Mk10 with .5in ranging machine gun; infra-red driving and fighting equipment

Other Variants

Vickers Modified Centurion In 1973 Vickers introduced a refitted Centurion complete with a new General Motors 720bhp diesel engine, semi-automatic transmission and improved gun-control equipment

ARV Centurion Mk2 (FV4006) Armoured Recovery Vehicle, a Mk3 fitted with a winch of a capacity of 91,445kg and armed with a 7.62mm machine gun

BARV Centurion (FV4018) Beach Armoured Recovery Vehicle, turretless and able to operate in water to a depth of 2.90m (9ft 6in)

AVRE Mk5 (FV4003) Armoured Vehicle Royal Engineers, armed with a 165mm demolition gun and equipped with a hydraulically-operated dozer blade mounted on the hull front

Bridgelayer Mk5 (FV4002) Utilises a Mk5 hull and can lay a 13.72m (45ft) single track bridge in two minutes

Above: A Centurion Mark 12, complete with infra-red searchlight and smoke launchers. The penultimate Mark in the Centurion series, the Mark 12 embodied the improvements that had taken place in over a decade of research and development.

contra-rotation to assist in target acquisition.

Meanwhile, developments in tank gunnery had proceeded apace. The L7 105mm gun, designed by the Royal Armament Research and Development Establishment, was introduced in 1959. The new gun, together with appliqué armour for increased protection of the glacis plate, was retrospectively fitted to the majority of Centurions in the British Army. By further limitation of operating space, a total stowage of 70 rounds was achieved – compared with 63 rounds on the Centurion Mark 7 and 65 on the Centurion Marks 3 and 5 – despite the L7's larger rounds.

To increase the effectiveness of the 105mm gun even further, a .5in ranging machine gun was mounted in the mantlet and aligned with the main armament. This made for a simple and accurate method of range-finding and eliminated the need for corrections due to factors such as trunnion tilt and crosswinds. In addition, a thermal sleeve was fitted to the main armament, reducing distortion of the hot gun barrel due to the uneven cooling effects of wind or rain. Infra-red equipment for night fighting and driving was also installed on later models including the Mark 13, the last production Mark of the Centurion series.

Centurion production ended in 1961 but the tank remained in service with the British Army until 1967, when it began to be replaced by the Chieftain MBT. Overseas, the Centurion's life has been extended further still. A mainstay of the Indian Army, the Centurion proved more than a match for the US-built Pakistani M47 and M48 tanks during the border conflicts of 1965, and in the war of 1971 the Centurion again outfought the Pakistanis' tank force – this time comprising the T59, the Chinese version of the Soviet T54.

The Australian Army sent a squadron of Centurions to Vietnam in 1968 where, in spite of the difficult terrain for tank warfare, they proved useful in an infantry support role; the tank's thick armour was able to survive the effects of North Vietnamese RPG rocket launchers and its canister rounds were very effective against unprotected infantry, as well as for clearing jungle and vegetation covering enemy bunkers. In the wars of the Middle East between Israel and the Arab states the Centurion achieved further renown when, in the hands of the Israeli Defence Force, it was acknowledged as one of the outstanding battlefield weapons of the 1967 Six-Day War and the Yom Kippur War of 1973. From a medium cruiser tank of the World War II period, the Centurion had come a long way.

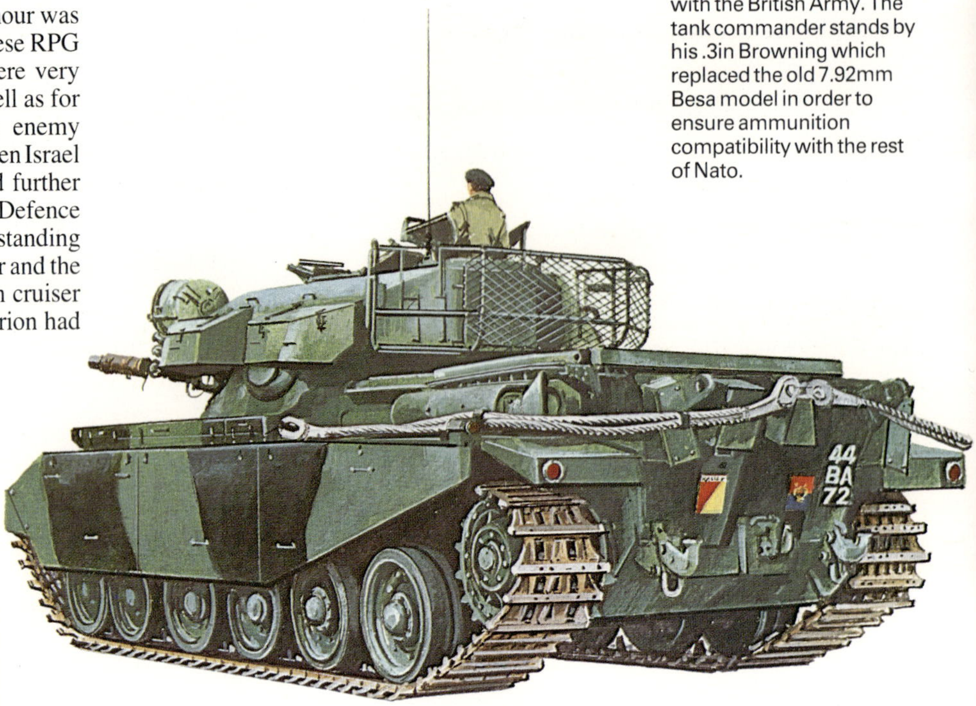

A Centurion Mark 13 – the final model to see service with the British Army. The tank commander stands by his .3in Browning which replaced the old 7.92mm Besa model in order to ensure ammunition compatibility with the rest of Nato.

Below: The .5in ranging gun and the .3in co-axial machine gun are situated within the turret mantlet on the right of the 105mm main gun. Above the mantlet is the infra-red search light, an invaluable aid to night-fighting.

Centurion Mark 13 MBT

Crew 4
Dimensions: Length (gun included) 9.85m (32ft 4in); width 3.39m (11ft 1½in); height 3.01m (9ft 10½in)
Weight Combat loaded 51,820kg (114,250lb)
Ground pressure 0.95kg/cm^2 (13.2lb/in^2)
Engine Rolls-Royce Meteor Mk IVB 12-cylinder, liquid-cooled petrol engine developing 650bhp at 2,550rpm

Performance Maximum road speed 34.6km/h (21.5mph); range (road) 190km (118 miles); vertical obstacle 0.91m (3ft); trench 3.55m (11ft); gradient 60 per cent; fording 1.45m (4ft 10in), with kit 2.74m (9ft)

Armour Min-max 17-152mm (0.67-6.08in)
Armament One 105mm L7A2 gun; one .5in ranging machine gun; two .3in machine guns – one co-axial with main armament, the other mounted on the commander's cupola; two six-barrelled smoke dischargers, one on each side of the turret

CHIEFTAIN

The Chieftain is currently in service with Iran, Iraq, Kuwait, Oman and the United Kingdom. Those in British service are being progressively modernised with the Improved Fire Control System (IFCS) and the Tank Observation and Gunnery System (TOGS). Firepower has been improved with the introduction of the new Royal Ordnance APFSDS round which, with its high muzzle velocity, can pierce the frontal armour of the latest Soviet tanks. In the British Army the Chieftain is now being supplemented by the Challenger, the first vehicles coming off the production line at Royal Ordnance Leeds in early 1983. The remainder of the British Army's Chieftains will probably be supplemented by a new MBT which is at present at the concept stage.

When proposals were first made in the early 1950s to replace the Centurion tank, which had entered service with the British Army in 1945, the question of the three conflicting priorities in the design of any armoured fighting vehicle again came under consideration. The order of these priorities agreed for the Centurion's successor, the Chieftain, remained as for its predecessor: firepower, protection and mobility.

Experience during World War II had underlined the importance of a high-performance gun; and the British considered the best anti-tank weapon to be the tank itself, with a powerful main armament capable of long-range target engagement. Next, protection was essential for the crew, and that had to include relative immunity from the effects of NBC (nuclear, biological, chemical) warfare as well as from shell-fire. Mobility was the third priority; restriction in mobility is the price inevitably paid for the insistence upon a high level of protection. A fourth and vital consideration – flexibility – is achieved in the Chieftain through its comprehensive Clansman radio communications system, which enables it to operate as a fast-reacting combat platform on the battlefield.

In 1956 Leyland Motors produced three prototype

Previous page: The crew of a Chieftain tank of the 17/21st Lancers, gun traversed in readiness, take a rest during training. Above: Chieftain main armament, the formidable 120mm gun. Below: An ARV (armoured recovery vehicle) variant winches a stranded vehicle back to safety. The ARV's capstan is capable of a pull of 30 tonnes.

Above: Two further variants – the Assault Vehicle Royal Engineers (AVRE) prototype which never saw production (top) and the Mark 5 ARV(P) with an ATLAS AK6000M crane developed for the Iranian Army. Above right: Combined tank and infantry exercises at Bovington. Below: A Chieftain Mark 3 has its L60 engine changed.

tanks designed along the lines laid down by the General Staff, and this led to the requirement being issued, two years later, for the FV4201 – the Chieftain Main Battle Tank. Designed by the Fighting Vehicle Research and Development Establishment, this new vehicle resembled the Centurion to some extent, but had a lower silhouette, a newly-designed turret without a mantlet, and a 120mm main gun. The lower silhouette was the result of positioning the centrally-sited driver in a reclining position (though only adopted when the tank is 'closed down'), while the three remaining crew members were accommodated in a turret whose profile afforded improved protection.

The P1 prototype appeared in 1959, to be followed by further prototypes which underwent an extensive series of trials, during which time engine and gearbox over-heating problems became apparent; extra strengthening of the transmission and lengthening of the engine compartment became necessary, unfortunately incurring additional weight. The hull was designed in three compartments consisting of cast sections welded together to form a driver's compartment, a central fighting compartment, and the rear

Left: A Chieftain FV4205 AVLB (armoured vehicle launched bridge) laying a No. 8 Bridge with a span of 22.8m (74ft 10in). Constructed of light alloy weighing only 12,000kg (26,896lb) the bridge can be laid in less than five minutes and when necessary can be recovered by the tank and reused.

engine section.

In May 1963 the Chieftain was accepted into British Army service and a number were provided for the 1st and 5th Royal Tank Regiments for troop trials in BAOR (British Army of the Rhine). In 1966 hot weather trials were successfully completed in Aden, and in 1967, in Germany, the first production Chieftains began to equip the 11th Hussars, the first regiment to receive them. Some 800 are currently employed in the British Army, the majority in nine regiments in BAOR and the remainder in two United Kingdom-based regiments. The tanks have been built at the Royal Ordnance Factory, Leeds (which has produced

Above: With turret-mounted searchlights blazing, the Chieftain's main armament is put to the test on the night firing ranges. Red lights indicate the use of live ammunition.

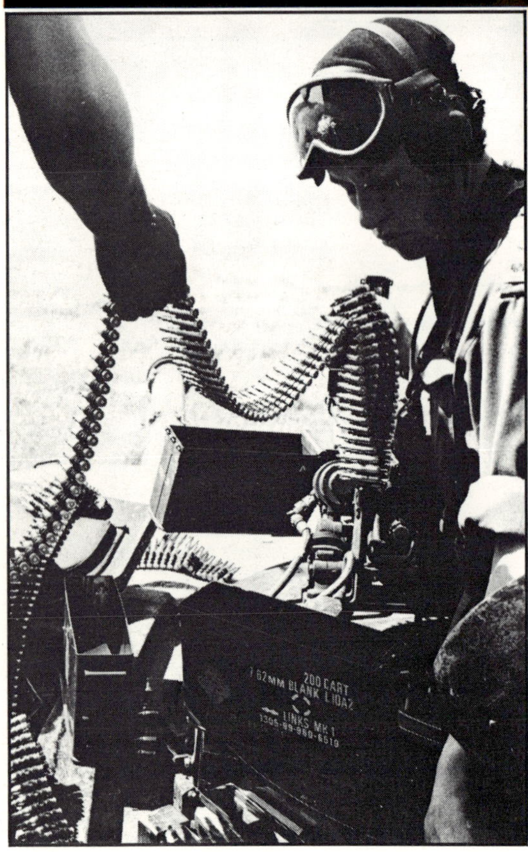

Left: Loading the Chieftain's cupola-mounted 7.62mm machine gun. The gun can be aimed and fired from within the cupola while maintaining NBC protection and automatically ceases firing when five rounds remain in the belt. This allows the commander to reload the gun on the battlefield without opening his hatch.

Chieftain Mark 5 MBT

Crew 4
Dimensions Length (gun included) 10.79m (35ft 5in); width 3.66m (12ft); height 2.89m (9ft 6in)
Weight Combat loaded 55,000kg (121,250lb)
Ground pressure 0.9kg/cm^2 (14.2lb/in^2)
Engine Leyland L60 Mk7A two-stroke multi-fuel six-cylinder engine with 12 opposed pistons developing 730bhp at 2250rpm

Performance Maximum road speed 48km/h (30mph); range (road) 450km (280 miles); vertical obstacle 0.91m (3ft); trench 3.15m (10ft 4in); gradient 60 per cent; fording 1.07m (3ft 6in)

Armour Conventional, 150mm (5.91in) maximum
Armament One 120mm L11A5 gun; one .5in ranging machine gun (retrospectively being replaced by a laser rangefinder); two 7.62mm machine guns – one co-axial with the main armament, the other mounted on the commander's cupola; two six-barrel smoke-grenade dischargers, one on each side of the turret

over 1900), and the Vickers factory at Elswick, Newcastle.

The 120mm L11A5 rifled tank gun provides the Chieftain's main armament. With the one-tonne barrel encased in a thermal sleeve the gun is well capable of engaging and destroying any potential opponent at long range. The latest barrels have an MRS (muzzle reference system) incorporating a mirror which enables the gunner to check the alignment of his sight with the barrel's axis from inside the vehicle, while ESR (electro-slag refined) steel barrels prolong bore life. The gun can elevate to plus 20 degrees and depress to minus 10 degrees.

The Chieftain's main types of ammunition are APDS-T (armour piercing discarding sabot tracer) which propels a tungsten core projectile at 1370 m/sec (4500 ft/sec) out to 3000m (3600 yds) and HESH (high explosive squash head), with slower muzzle velocity of 670 m/sec (2200 ft/sec) and a range of 8000m (8750 yds). Bagged charges – which save space in turret confines – are stored below turret-ring height in coolant-filled charge bins to minimise fire risk. Other ammunition types include canister, illuminating and smoke, while APFSDS (armour piercing fin stabilised discarding sabot) has recently been introduced into the British Army.

The gunner and loader can provide a rate of fire of up to 10 rounds per minute for the first minute; thereafter a rate of six can be maintained. A Marconi-designed stabilised gun control system and various Marks of tank laser sight – now standard on all British Army Chieftains – allow the gun to remain stable while the tank is on the move and the target to be tracked through a ballistic graticule pattern and laser-ranged by gunner or commander to an accuracy of 10m (11 yds) even at the longest ranges.

The Marconi Space and Defence System's IFCS (integrated fire-control system) – at present being retro-fitted to earlier British Chieftains – takes into account (via its computer) vehicle attitude, weather conditions, range, aim-off, and even fling (the upward flip of the projectile leaving the individual barrel), all within a few seconds, and offers a systems accuracy of 90 per cent to a range of 5000m (5470 yds).

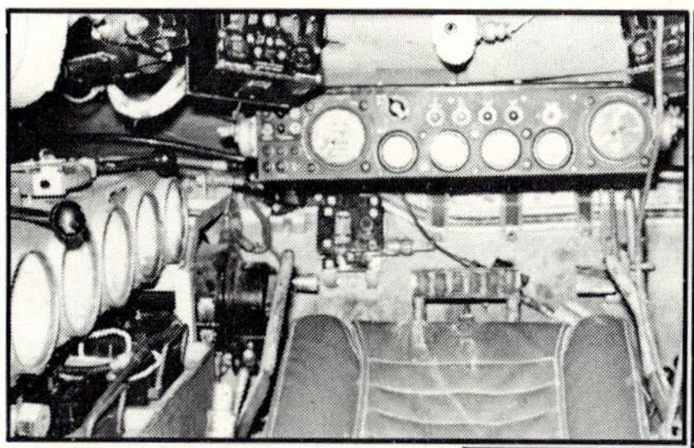

Above: The driver's position, surrounded by batteries and main gun projectile stowage bins, is set in the front of the Chieftain's hull. When the tank is 'closed down' the driver lies prone but is provided with a driving hatch for use when not in combat.

Above: The gunner's position is situated in the right of the Chieftain's turret. The retrofitted fire control system which includes laser range-finding equipment replaces the .5in ranging gun and virtually guarantees a first round hit on a stationary target at 3000m (3281 yds).

This system has dispensed with the 0.5in Browning ranging gun of earlier Marks of the Chieftain, but secondary armament remains in the two 7.62mm machine guns, one on the commander's cupola and one mounted co-axially with the main armament. The turret, which can traverse through 360 degrees in 16 seconds, has smoke grenade dischargers fitted on each side.

The Chieftain's high level of protection is provided by a combination of armour thickness, and turret and hull design. The sloped profile of the cast turret and the emphasis on frontal protection both there and on the glacis plate primarily ensure protection against an enemy attack to the front, but the shallow 'V' hull profile and skirting plates for the track tops also play a part in the overall protection scheme.

A three-stage NBC pack attached to the turret rear provides protection for the crew and obviates the need for respirators by providing air filtration, or ventilation in a non-NBC environment. The crew can continue to operate closed-down for 48 hours provided pressure inside the tank is maintained. On earlier

Below: Obscuration, the blinding effect of dust thrown up by muzzle blast, is one of the problems of the Chieftain's 120mm main armament. It can, however, provide effective cover for rapid movement in very dry or desert conditions.

Marks an over-pressure system was used, but this is being replaced by a full filtration pack. IR (infra-red) detectors are fitted to the turret top to provide audible warning of enemy IR emissions.

A Nato requirement introduced during the design stage of the Chieftain called for a multi-fuel engine which led to a redesigning of the engine compartment, and for the first decade of the tank's service with the British Army its engine was bedevilled with problems. Piston rings fractured, cracks and displacement occurred in rear gear case and cylinder liners, and cooling and air filter inadequacies became apparent. The early Leyland L60 engine – a 12-cylinder two-stroke design – in the Mark 1 version provided a mere 585bhp, which fell considerably short of the specified figure of 750. The Mark 2 improved this to 650bhp at 2100rpm, but an effective power-to-weight ratio was not achieved until the Mark 8A was introduced. This engine had an output of 750bhp, and its improved reliability made it the first really satisfactory power unit.

Drive from the engine leads via the centrifugal clutch to the TN 12 epicyclic gearbox with a Merritt-Wilson differential steering system and electro-hydraulic gear selector. There are six forward and two reverse gears.

The suspension is of the Horstmann type and consists of three units on each side of the tank which have dual road wheels. Three sets of horizontal springs are fitted, and the drive sprocket is at the rear. The tracks are of cast manganese steel with detachable rubber pads which minimise damage to road surfaces in peacetime.

The Chieftain is fitted with a comprehensive range of optical equipment. The driver's periscope can be replaced by a passive image intensification night-driving sight which is undetectable and considerably improves the tank's chance of survival on the night battlefield. The loader, too, has a periscope. The commander is provided with 360 degree vision by nine episcopes, and an additional sight which enables him to override the gunner if necessary. An infra-red white-light searchlight for target illumination is fitted to the turret left-hand side but its active emissions are, of course, detectable.

The Chieftain has run to 12 Marks in British Army service, of which the most numerous are the Mark 3/3 and Mark 5; the latter Mark upped the 3/3's main armament ammunition storage capacity from 53 to 64 rounds in addition to other improvements to the engine and gearbox. Marks 9 to 12 are all IFCS-fitted versions. The Chieftain Mark 5 provided the basis for the ARV (armoured recovery vehicle) equipped with double capstan winches and dozer blade, and the AVLB (armoured vehicle launched bridge) provides the means of transportation for two types of tank bridge, which can span any gap between 12 and 22m (13 to 24yds).

Above: A pair of front-line Chieftains operating in Germany where current Warsaw Pact numerical superiority in AFVs emphasises the crucial need for quick response and superior combat skill.

Below: The low profile of a Chieftain MBT emerges stealthily from the woods while on exercise with the British Army of the Rhine.

From Chieftain to Challenger

Since its introduction into British Army service in the late 1960s the Chieftain has been acknowledged as one of the world's most powerful battle tanks. Heavily armoured and armed with a large and highly accurate 120mm main gun, the Chieftain has been well suited to its role as armoured guardian of Britain's forces in northern Germany. During the 1970s the tank's extensive engine and transmission problems were largely solved, and over a course of 12 Marks a comprehensive series of improvements ensured that the Chieftain would still be able to hold its own well into the next decade.

While these developments were taking place, plans were being drawn up for a successor to the Chieftain. Initially, ideas were put forward for an international cooperative venture but this eventually come to nothing and Britain went ahead with its own project, the MBT-80. The new project did not come to fruition, however, for two main reasons: firstly, the cost of the whole undertaking rose dramatically and, secondly, the drawn-out schedule of the MBT-80 programme meant that a new tank would not be ready to take the field by the mid-1980s, the date when it was considered that the Chieftain would be ready for replacement. Dramatic advances in Soviet tank technology which saw the introduction of the T64/T72 series in the 1970s made the need for a speedily updated MBT all the more imperative.

The MBT-80 programme was abandoned in 1980 in favour of a faster solution albeit at the cost of a less advanced design. The solution lay in the Chieftain itself. Whereas the British Army Chieftains had undergone a long series of improvements, they were in the main of a relatively minor nature; but those Chieftains destined for the export market had been radically overhauled and consequently they formed the basis for the FV4030/4 Challenger.

A number of countries in the Middle East had shown considerable interest in the hard-hitting Chieftain and orders were placed by Iran, Jordan, Oman and Kuwait. Iran was the largest purchaser: in 1971 orders were place for the impressive total of 750

Although the Challenger with its improved powerplant and Chobham armour is now supplementing the Chieftain in service with BAOR, Chieftains such as these of the Queen's Royal Irish Hussars remain the backbone of the British armoured forces in Germany. Left: A Chieftain of the 3rd Royal Tank Regiment ploughs its way through a dust bowl at Fallingbostell in Germany.

Mark 3/3(P) and Mark 5/5(P) tanks. These were delivered by 1978, to become the spearhead of Iran's armoured formations. The Shah of Iran's modernisation programme for his country was centred on the armed forces, and through its massive oil revenues Iran was able to afford the most advanced weapons from the West. In late 1974 Iran placed a further order worth £500 million for 193 improved Chieftains, 125 Shir (Lion) Iran 1s and 1225 Shir Iran 2s. The Shir Iran 1 was based on the Chieftain but had an uprated power pack and the Shir 2 was an even more advanced model that made use of the latest Chobham armour, a revolutionary armour type thought to be constructed of layers of aluminium, titanium, ceramics and other unspecified materials. The delivery of the improved Chieftains had taken place and the production of Shir 1 was nearing completion when, in February 1979, the overthrow of the Shah's regime led to the ultimate cancellation of the remaining orders.

Meanwhile, in 1975, Kuwait had ordered 143 Vickers-built Mark 5/5(K) versions of the Chieftain and in November 1978 Jordan placed an order for 278 Khalid MBTs. The Khalid was in essence the Shir 1, equipped with a 12-cylinder Rolls-Royce engine, Marconi IFCS and a Pilkington PE Condor day/passive night sight. Deliveries commenced in mid-1981, and in the same year Oman received a small number of ex-British Army Chieftains.

The Shir 2 order was utilised by the British Army, being designated the FV4030. A number of prototypes were developed and in July 1980 an order for 243 tanks was announced. Known as the Challenger, the first prototype FV4030/4 emerged in late 1981 and made its public appearance the following year.

The most important feature of the Challenger is its 12-cylinder Rolls-Royce diesel engine, capable of producing 1200bhp at 2300rpm. This will give the tank a far more impressive performance, more in line with its American and German counterparts. In order to exploit this increased power a new hydropneumatic suspension system has been fitted independently on each wheel, thereby increasing mobility considerably. In line with these improvements, a TN37 transmission with torque converter has been fitted, as has a new Howden cooling system.

While the 120mm L11A5 rifled gun has been retained, it is probable that it will be replaced by a 'modern technology' main armament which will have an improved performance and will utilise a new ammunition system including an APFSDS round. Chobham armour is used, which will make the Challenger virtually safe from most chemical anti-tank weapons such as HEAT and HESH, and consequently will provide the MBT with a new lease of life on the European battlefield.

As only 307 Challenger MBTs have been ordered they will, in the first instance, augment rather than replace the Chieftains in BAOR. Inevitably, though, the Chieftain will be gradually phased out of service as more Challengers are produced, or when a radically new MBT is manufactured.

Below: A prototype of the new Challenger MBT on trials in England. Based on the Shir 2 and modified to meet British Army requirements, the first Challenger unit was formed in West Germany in 1983. The Challenger incorporates the Shir 2 and Khalid's 1200hp CV12 TCA diesel, a radical improvement on the troublesome Leyland L60 engine fitted to earlier Chieftains. Its increased power, Chobham armour protection and NBC fit make the Challenger a formidable opponent on the modern battlefield.

LEOPARD 2

The Leopard 2 is the most advanced tank in service with the West German Army, which expects to take delivery of 1800 of them by 1987. The Netherlands have ordered 445 Leopard 2s, the first of these entering service in 1982. In August 1983 the Swiss announced that they intended to order 420 Leopard 2s, the first 35 coming from West Germany, the remainder being made under licence in Switzerland. The Leopard 2 has outstanding armour protection, mobility and firepower, the latter being due to the Rheinmetall 120mm smooth-bore gun which has also been selected to arm the American Abrams M1A1 MBT. Currently under development are armoured recovery and armoured engineer versions of the Leopard 2.

The development of the German Leopard 2 MBT resulted from the abortive collaboration by America and West Germany during the 1960s on a project to build a standard new main battle tank designated the MBT-70. Due to escalating costs the project was abandoned in 1970 but the Germans proceeded to develop a vehicle incorporating a number of features of the MBT-70 programme, including the power-plant. In 1977 the Bundeswehr ordered 1800 Leopard 2s from contractors Krauss-Maffei of Munich and MaK of Kiel and the first production models were delivered in 1979. The Leopard 2 is also in service with the Dutch armed forces who placed an order for 445 tanks in 1979, and in 1983 it was accepted by the Swiss Army as a replacement for its Centurions and Pz61s.

In the context of operations on the Nato Central Front, for which the Leopard 2 is designed, advances in Soviet tank design dictated the need for major improvements to the capabilities of the Leopard 1. Developments in automotive technology during the 1970s allowed tank designers more weight without sacrificing mobility and the compromise between firepower, mobility and protection achieved in the Leopard 1 became obsolescent. One of the prime considerations in the design of the Leopard 1 was its battlefield mobility and in order to maintain this capability the Leopard 2 was provided with a turbocharged multi-fuel engine which developed 1500hp as opposed to the Leopard 1's 830hp power-pack. Track width was also increased from 550mm (21.65in) to 635mm (25in) which, combined with advanced torsion-bar suspension, provided the Leopard 2 with as high a road speed as its predecessor and excellent cross-country agility despite the 15 tonne increase in combat weight.

In order to meet the new requirements for protection demanded by advances in anti-tank ammunition the Leopard 2 was to be fitted with spaced armour. Rounds such as HEAT and HESH use chemical energy to penetrate armour by detonating an explosive charge directly against the target plate but are only effective against a single thickness of armour. A sandwich of armour, with a space between the inner and outer plates, prevents the hot gases and fragments which are blasted through the outer plate from penetrating the main inner armour plate and reaching the interior of the vehicle. Prototypes of the Leopard 2 were provided with spaced armour but at this stage of development the new British concept of Chobham armour came onto the market. Further collaboration between America and West Germany in the mid 1970s led to the production of a Leopard 2 AV (austere version) which was calculated to meet American cost and performance requirements in 1976. This opportunity to alter the original concept allowed German tank designers to incorporate Chobham armour into the design and the Leopard 2's armour is believed to be a combination of spaced and Chobham, providing a high level of immunity against both chemical projectiles and straightforward armour-piercing ammunition.

As tank armour has become more sophisticated and has considerably reduced the efficacy of chemical

Previous page: A Bundeswehr Leopard 2 on manoeuvres in northern Germany. The Leopard 2, alongside the American Abrams and the British Challenger, provides a formidable combination of firepower, mobility and survivability for the Nato front-line arsenal. Above: While its predecessor the Leopard 1 sacrificed weight in favour of mobility, the 55 tonne Leopard 2 is fitted with a rugged 1500hp turbocharged powerpack, allowing its designers to concentrate on increasing the tank's armour while retaining high performance and cross-country agility.

rounds, so tank guns have been uprated to provide a much higher velocity for kinetic energy armour-piercing shot. Although the Leopard 2 AV retained the rifled 105mm main armament at the Americans' request, the Bundeswehr opted for the greater fire-power of the 120mm Rheinmetall smoothbore gun which was fitted to the Leopard 2. Two main types of ammunition are deployed – HEAT-MP-T (high-explosive anti-tank multi-purpose tracer) and APFSDS-T (armour-piercing fin-stabilised discarding-sabot tracer). An unusual feature of the Rheinmetall ammunition is the combustible cartridge case; this eliminates the problems of spent shell-case ejection since only the base stub remains after a round has been fired and this is easily collected in a bag located under the breech. The 120mm gun is fully stabilised and both the gunner and the commander are provided with stabilised sights, allowing them to lay and fire the main armament on the move with a high first-round hit probability. The gunner is also provided with a laser range finder and thermal image unit linked to the fire-control system. Passive night-vision equipment and NBC systems are standard.

Secondary armament is similar to the Leopard 1 and consists of two 7.62mm machine guns, one mounted co-axially with the main armament and the other on the turret roof on either the commander's or loader's hatch for use against air attack. Eight smoke dischargers are fitted on each side of the turret.

The new generation of Nato MBTs, the American M1, the British Challenger and the Leopard 2 have

Top, left and right, and above: The Leopard 2's ability to survive armour-piercing and chemical anti-tank weapons depends on a combination of spaced and Chobham armour coupled with a good ballistic shape. Left: The squat turret provides a low target silhouette, while the armour arrangement gives a good protection against HEAT and HESH projectiles.

Left: A camouflaged Leopard 2 moves through wooded country on manoeuvres with the Bundeswehr.

Right: The Leopard 2's main armament is put to the test on the firing ranges. Despite the proven combat record of the British 105mm L7 series tank gun, the Leopard 2 was fitted with the heavier 120mm Rheinmetall smoothbore gun firing two types of fin-stabilised ammunition.

Below: With its main gun and turret well camouflaged, a Leopard 2 takes up a firing position behind scrub in open country. A 7.62mm machine gun is mounted on the loader's hatch while 16 smoke dischargers are fitted to the turret bustle.

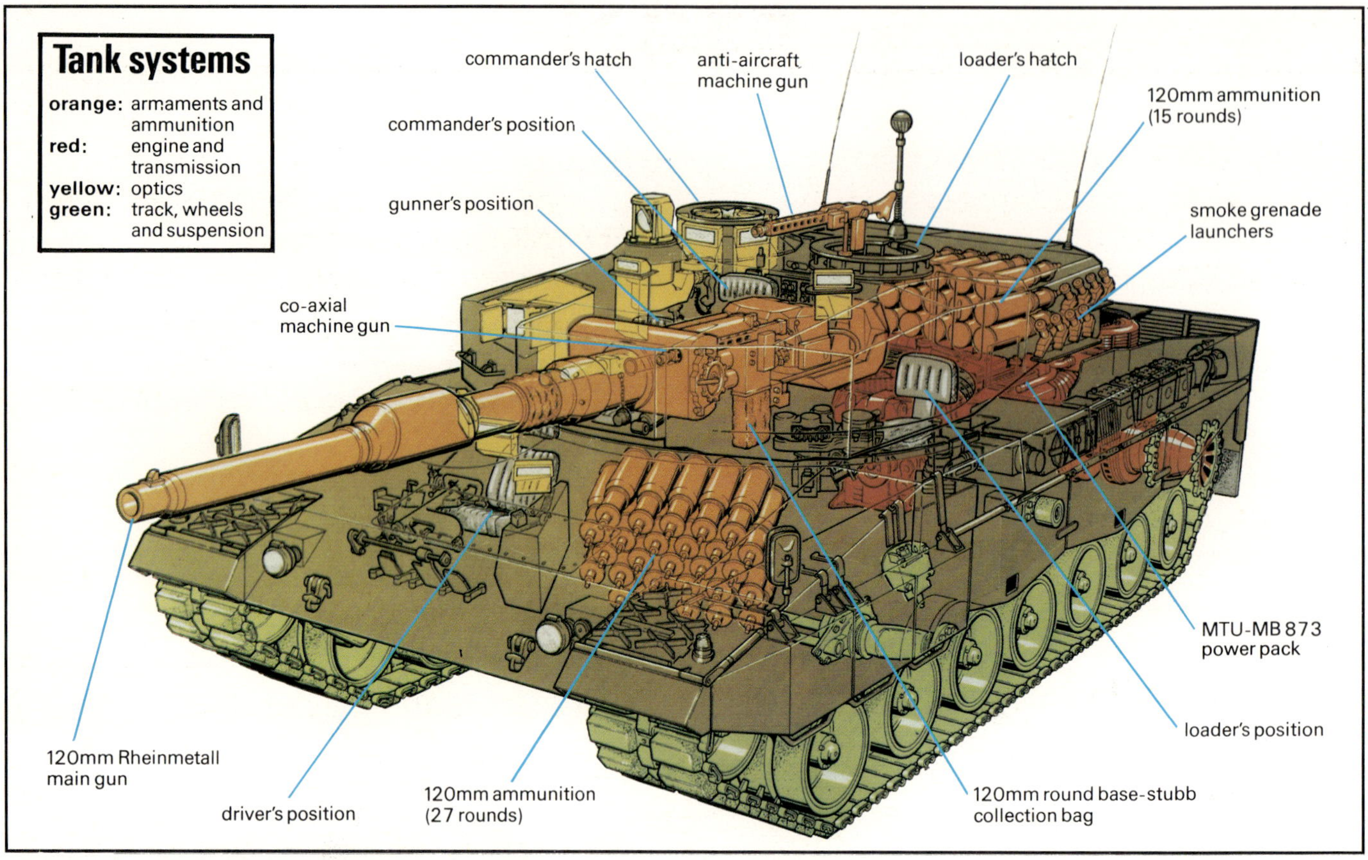

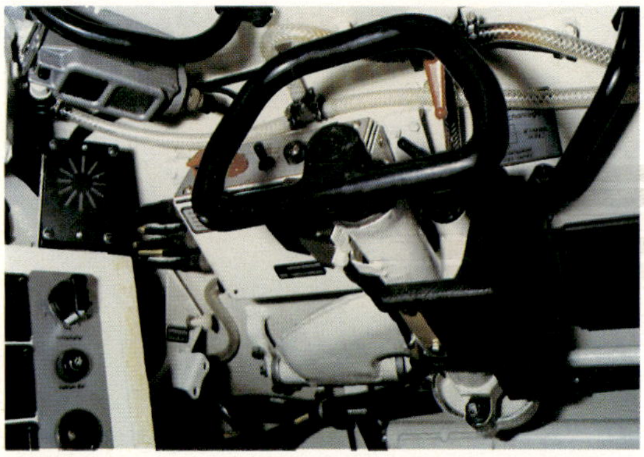

Left: The driver's position on the Leopard 2. In contrast to the Abrams the Leopard 2 is fitted with a steering wheel rather than a T-bar control. Below: A Leopard 2 turret being installed on the tank assembly line at the MaK works in Kiel. 810 vehicles were ordered by the German Army from MaK in 1977 with final deliveries of the Leopard 2 scheduled for 1986.

been criticised for not showing any really radical advance on the capabilities of their predecessors. It is true enough that the Challenger is an obvious descendant of the Chieftain, just as the Leopard 2 is heir to the Leopard 1, but both are considerably improved and more balanced machines with a more equal distribution of the cardinal tank virtues of mobility, firepower and protection than was possible to envisage in the 1950s when the original Chieftain and Leopard concepts were born. The 120mm Rheinmetall rounds can penetrate a standard Nato heavy tank target at some 2200m (2400yds) and the Leopard 2 is considered more than a match for any contemporary MBT.

Leopard 2 Main Battle Tank

Crew 4
Dimensions Length (gun included) 9.61m (31ft 6in); width 3.7m (12ft 1½in); height 2.79m (9ft 2in)
Weight Combat loaded 55,000kg (121,220lb)
Engine MTU MB 873 12-cylinder multi-fuel turbocharged engine developing 1500hp at 2600rpm

Performance Maximum road speed 72km/h (45mph); range (road) 550km (340 miles); vertical obstacle 1.1m (3ft 7in); trench 3m (9ft 10in); gradient 60 per cent; fording without preparation 0.8m (2ft 7½in), with preparation 2.35m (7ft 8in), with snorkel 4m (13ft 1in)

Armour Classified but believed to be a combination of spaced and Chobham
Armament One 120mm Rheinmetall smoothbore gun; one 7.62mm machine gun co-axial with main armament; one 7.62mm machine gun mounted on the commander's or loader's hatch; eight smoke dischargers on each side of the turret

M48

The M48 was in production between 1951 and 1959, in which time 11,703 were built by the Chrysler Corporation, Ford Motor Company, the Fisher Body Division of General Motors Corporation and Alco Products. In 1985 it remained in front-line service with some 20 countries all over the world, the largest users being Israel and the United States. Many countries have replaced the original petrol engine with a diesel (for greater operational range) and the 90mm gun with the 105mm M68, which is essentially the famous British L7 made under licence in the United States. Further development of the M48 resulted in the M60 series of MBT, which is still being produced at the American Detroit Tank Plant to supply the export market.

The M48 Patton Medium Tank entered service with the United States Army in 1953. In December 1950 the Chrysler Corporation had received a contract to design and build a medium tank to replace the transitional T47, and one year later the prototype was ready. Production began in 1952, but the first years of the M48's service life were beset with problems. Reliability proved uncertain and mechanical breakdowns severely limited the tank's efficiency. The stereoscopic range finders also caused difficulties. These defects were, in time, remedied and modifications have led to the present-day M48A5 version.

The original M48 had cupola fitting problems so that eventually a cupola with the cradle-mounted .50in M2 machine gun was installed. Turret and hull were of cast steel construction and the latter was divided into driving, fighting and engine compartments. Suspension was of torsion bar type and there were six road wheels each side with five track return rollers. A Continental AVDS-1790-7C 12-cylinder petrol engine powered the 47-tonne tank. In the original version range was limited and while the provision of additional fuel tanks provided a greater radius of action they also increased vulnerability. The M47's fire control equipment was retained in the M48.

The M48A1 mounted an improved cupola (the M1) with five periscopes and gunsight. Its .50in machine gun had a plus 60 degree elevation and minus 10 degree depression, and thus had an anti-aircraft as well as a ground role. Main armament was the 90mm M41 gun with distinctive bore evacuator and blast deflector. The gun had automatic ejection and a

Previous page: The United States Army has been well served by the M48 series of tanks, which first came into service in 1953, and despite the introduction of the M1 Abrams MBT they continue to provide battle-front support for America's armoured divisions. Above: An Israeli M48A2 advances across the desert in the face of Egyptian anti-tank fire during the Six-Day War.

Right: The M48 was widely used during the Vietnam conflict and was considered a success by its crews – to the surprise of some tacticians who failed to realise the virtues of tanks in modern jungle warfare, despite problems of terrain and deployment. US Marine Corps tanks prepare for action (top and centre), and an M48A3 advances against enemy positions during Operation Sitting Duck in 1966 (bottom).

Above: The crew of an Israeli M48A2 emerge from their turret after an action during the Six-Day War. Alongside the Centurion the M48 was the major armoured success of the 1967 conflict.

vertically sliding breech, and the gunner occupied the front right-hand side of the turret in front of the commander, with the loader on the left. Sixty rounds of fixed case ammunition were available and types included HE, AP, HVAP and HEAT. A co-axial .30in M1919 machine gun was originally fitted as secondary armament but was later replaced by a 7.62mm M73. For battlefield illumination an 18in 2000 watt searchlight was carried, later to be replaced by a Xenon infra red/white light type. A tensioning idler was installed in front of the rear drive sprocket to counteract the tendency to shed tracks, and on later types two of the five track return rollers were removed. The AVDS-1790-7C petrol engine was retained, though fuel injection was introduced.

In the mid-1950s the M48A2 appeared. This had a larger M19 cupola with 360 degree traverse, and the engine compartment was completely redesigned and improvements were also made to the commander's override ability over the 90mm gun. Fuel capacity was increased and the transmission – which was still proving unreliable – received attention. The track compensating idler spindle was redesigned and the track tensioning idler of the M48A1 removed. Overheating problems attributable to the engine cooling fan resulted in improvements being made to the cooling system. Later versions carried the prototype turret for the M60 and had multi-fuel engines.

Armed with a 90mm main gun an M48A2 pauses for a mechanical inspection during the Israeli advance against Arab positions in 1967. A considerable weapon in its own right the M48's success, nonetheless, owed much to the tactical ability of the Israeli tank crews who consistently outfought their Arab opponents.

The Israelis fitted British 105mm L7 guns to their M48A2s which they received in the 1960s, and made a number of modifications including fitting diesel engines. Both the Israelis and the Jordanians used the tank in the 1967 war. The Israelis found it faster and more reliable than the Centurion although less well armoured, and by 1973 the M48 (and the new M60) had begun to supplement the Centurion in Israeli front-line service.

In 1964 the M48A3 introduced Continental diesel engines into the M48 series with the AVDS-1790-2A. Fuel capacity was increased and range improved. The driver had three M27 periscopes, the middle one of which could be replaced by an M24 infra red scope. The commander's M1 cupola had five vision blocks and an M28C sight for the .50in machine gun. A coincidence rangefinder with x10 magnification and 4400m (4800yds) range was installed. The gunner had both periscope and telescope with x8 magnification. No fire warning system was fitted, but a floor escape hatch enabled the crew to vacate the vehicle under its own cover. An M8 series dozer blade was available. In 1967 nearly 600 M48A1 tanks were brought up to M48A3 standard.

The M48A3 was the standard American tank in Vietnam and saw service there between 1965 and 1973. The complement of track return rollers was again made up to five, and cutting bars fitted to the tank's front enabled it to clear dense vegetation. Additional machine guns were mounted to improve firepower in close country. The tank's main armament proved particularly useful during combined infantry/armoured operations and the 90mm canister and beehive rounds were used to good effect. Some problems were understandably encountered with engine overheating and mud clogging the tracks, but the

Below left: An Israeli M48 prepares to pass its leader – whose tank has hit a mine – at one of the crucial actions of the Yom Kippur War on the west bank of the Suez Canal. Below: M48s were used by the Jordanians as well as the Israelis during the Six-Day War. Here a unit of Israeli Shermans pass a knocked-out Jordanian M48 near Kabatiya in Samaria. Bottom: An M48A2 races across the Sinai Desert in the Six-Day War, while in the background Egyptian armoured vehicles burn and explode.

Some of the differences in turret layout can be seen between an Israeli M48A2 (above) which has been fitted with a 105mm main armament and a standard M48A2 (above right) with its less powerful 90mm gun. Right: An M48A3 supports infantry operations on a beach assault by men of the 7th US Marines.

Below: A US Marine M48A3 takes part in Operation Arcadio in 1966. The M48 proved popular with both the Marines and the army.

M48 was well liked by its users and despite adverse conditions the tank performed very well.

The M48A4 was essentially an M48A3 with the M60 turret and 105mm L51 gun. The old 90mm gun was deemed to be inadequate to counter the main armament of contemporary Soviet tanks and so the M48's firepower was updated. The M48A5 retained the 105mm gun and added a co-axial 7.62mm machine gun, while the loader received an additional machine gun. New tracks were fitted and a coincidence rangefinder installed.

In the M48 family were to be found the C series of mild-steel lightweight training tanks, AVLB scissors bridgelayers capable of crossing an 18.2m (60ft) gap, the M67 series of flamethrowing tanks, the M88 ARV and mine clearance variants.

The M48 was intended to equip the United States Army in the Korean War, though it arrived too late to see action in that conflict. It has since seen extensive service in a number of theatres of war, including the Indo-Pakistan war in 1965, where its Pakistani operators mishandled the tank and contrived to lose it in large numbers. Recipients have included West Germany (where at present the Bundeswehr operates 650 M48-series tanks rebuilt to house the 105mm L7 gun), Greece, Iran, Israel, Jordan, Morocco, Norway, Pakistan, South Korea, Spain, Taiwan, Thailand, Turkey and South Vietnam. Over 11,000 versions have been built, and in its upgunned form the M48 remains a useful medium tank into the 1980s.

Modified Israeli M48s take part in exercises shortly after the end of the Six-Day War. A Centurion is just visible in the background behind the lead M48.

M48A3 Medium Tank

Crew 4
Dimensions Length (gun included) 7.44m (24ft 5in); width 3.63m (11ft 11in); height (including cupola) 3.12m (10ft 3in)
Weight Combat loaded 47,173kg (104,000lb)
Ground Pressure 0.83kg/cm^2 (11.8lb/in^2)
Engine Continental AVDS-1790-2A 12-cylinder air-cooled diesel engine developing 750bhp at 2400rpm

Performance Maximum road speed 48km/h (30mph); range (road) 463km (288 miles); vertical obstacle 0.92m (3ft); trench 2.59m (8ft 6in); Gradient 60 per cent; fording 1.22m (4ft), with kit 2.44m (8ft)

Armour Min-max 12.7mm-120mm (0.50-4.80in)
Armament One 90mm M41 gun; one 0.3in M1919A-4E1 machine gun co-axial with the main armament; one 0.5in machine gun in commander's cupola

Right: An M48A1 climbs a steep gradient. Distinguishing features of this variant are the commander's cupola, the flat engine decks and the undercut rear hull plates.
Below: The M48 AVLB (armoured vehicle launched bridge) sees serious action during fighting in South Vietnam in 1966.

M60

Although the Abrams M1 MBT is entering service with the US Army in ever-increasing numbers, the M60 will remain the most common tank in US service until the 1990s. By 1984 over 15,000 M60 series had been built at the Detroit Tank Plant operated by the Land Systems Division of General Dynamics. Production is now for export only, the last production M60A3s having been delivered to the US Army several years ago. Most US Army M60A1s are now being upgraded to the M60A3 configuration to include a laser rangefinder, passive night vision equipment, new co-axial machine gun, British-type smoke dischargers and fully stabilised 105mm gun.

Previous page: Here two M60A1 tanks are on a live-fire exercise. Above: While observing the effects of gunfire the commander of an Israeli M60 ensures that his tank is well sited in a hull-down position.

The M60 was a development of the M48-series tank and was intended to replace medium and heavy classes of tank in US Army service with a main battle tank. The Chrysler Corporation initially received the production contract in 1959 for 180 examples. The first tanks were built in 1960 and entered service with the United States Army in November that year. The combination of gun and chassis proved successful and an order was placed for a further 770 M60s.

The year 1962 saw the introduction of the M60A1 variant. This improved an already sound design and featured an enlarged and pointed cast-armour turret with M19 cupola providing eight periscopes, as well as hull design changes to improve protection. The M60A1 was built with a cast hull, and was provided with torsion bar suspension with six road wheels. The drive sprocket was positioned at the rear, the idler at the front, and three return rollers were fitted. The Continental AVDS-1790-2A diesel engine developed 750bhp at 2400rpm. An NBC system was made available, with full infra-red/white-light equipment standard. The driver's optics were similar to those in the M48, but a passive night-vision sight replaced the M24.

The well proven 105mm gun with power-assisted elevation had 60 rounds available in the M60 and a further three in the M60A1. Its eccentric fume extractor mounted towards the rear of the barrel provided a recognition feature to distinguish the M60 from the M48. A rate of fire of between six and eight rounds per minute was attainable. The gun can be elevated to plus 20 degrees and depressed to minus 10 degrees and a 7.62mm M73 machine gun is co-axially mounted. The commander's M19 cupola has an electrically operated .50in Browning M85 machine gun which, with elevation of plus 60 degrees, has a dual role. For the .50in machine gun, 1040 rounds are available, and for the 7.62mm 5500. The M17C coincidence rangefinder provides a high level of accuracy up to 4400m (4800yds) and an electronic ballistic computer increases the chances of a first-time hit.

Right: An M60A1 of the US Army. Despite the attempt at camouflage the high turret silhouette of the M60 is only too apparent, a fault accentuated by the commander's cupola.

Right: Armed with a 105mm M68 gun an M60A1 blasts away on the firing range. Some 63 rounds of main armament ammunition can be carried within the tank.

Above and inset: The M60A2 incorporated a 152mm gun/launcher system, capable of firing a number of conventional ammunition types as well as the Shillelagh missile.

The M60A2 was developed in the mid-1960s and introduced a new concept in tank gunnery. The idea was to mount, in a redesigned turret, a 152mm weapon capable of either launching a Shillelagh missile or firing conventional rounds with combustible cases. When firing the weapon in the missile mode the gunner acquires his target with infra red, then launches the Shillelagh which flies towards the target steered by command guidance. The disadvantage is that the M60A2 can only fire the missile – of which 13 are carried – from the halt; and stabilisation problems with the turret and tracking systems have caused severe difficulties. The complicated nature of a breech designed to accept a missile or a round, and the problems caused by incomplete combustion of the case, ensured a long delay in introducing the M60A2

RANGE
46

Left: The M60A1 was the major production model in the M60. Below left: The crew of an M60A1 take a rest while on an armoured exercise in central Norway. This M60 is fitted with a dozer blade.

Below: A column of Israeli M60A1s ploughs through the sands of the Sinai Desert. Bottom: A derivative of the M60A1, the M728 Combat Engineer Vehicle was armed with a 165mm demolition gun and, in addition, was equipped with a dozer blade, an A-frame boom with a 2.4m (8ft) lift and a winch with an 11-tonne pull.

into front-line service. In 1971, however, these problems were largely solved and during the 1970s a total of 526 Shillelagh-armed tanks entered US Army service.

Production continued with an up-rated M60A1, with hydraulic gun stabilisation and an improved air cleaner which led to the M60A3 in 1978. This version has an AN/VVG-2 laser rangefinder and XM21 ballistic computer which, linked to the all-weather capability AN/VGS-2 tank thermal sight in conjunction with a new APFSDS M735A1 Stabiloy round, provides a high degree of accuracy up to a range of 3000m (3300 yds). A laser fire-control system and muzzle reference system for the gunner ensure a good first-time hit probability rate. Many other improvements to optics, suspension and electrical systems have been made and further ones are planned to help keep the M60 abreast of contemporary developments.

The M60A3 arrived in Europe in mid-1979 and there were plans to build over 2500 of them. Israel and Jordan have both placed orders for this variant, and the US Marines operate it too. By early 1975 over 3000 M60 or M60A1 main battle tanks had entered US Army service, over 100 were with the Marines, and over 900 had been exported. By the time production was run down in 1982 more than 13,000 tanks had been built.

The M60 has proved to be a reliable vehicle in service and, while production of the advanced M1 Abrams tank is under way, the series will continue to provide armoured support in Europe well into the 1990s. During the war in Vietnam the M728 Combat Engineer Vehicle and AVLB versions saw extensive action; and the Middle East War of 1973 gave the M60 an opportunity to be tested in battle, where it performed creditably. An undoubted export success, it has been sold to many countries including Austria, Egypt, Ethiopia, Iran, Israel, Italy, Jordan, North Yemen, Oman, Saudi Arabia, Singapore, Somalia, South Korea, Sudan and Tunisia.

M60 MBT

Dimensions Length (gun included) 9.31m (30ft 6in); width 3.63m (11ft 11in); height 3.26m (10ft 8in)
Weight Combat loaded 48,987kg (108,000lb)
Ground Pressure 0.79kg/cm^2 (11.241lb/in^2)
Engine Continental AVDS-1790-2A 12-cylinder diesel engine developing 750bhp at 2400rpm

Performance Maximum road speed 48km/h (30mph); range (road) 500km (310 miles); vertical obstacle 0.91m (3ft); trench 2.59m (8ft 6in); gradient 60 per cent; fording 1.22m (4ft), with kit 2.44m (8ft)

Armour Min-max 12.7mm-120mm (0.50-4.80in)
Armament One 105mm M68 gun; one 7.62mm machine gun co-axial with main armament; one 0.5in machine gun in commander's cupola

Top: A unit of Israeli M60A1s halts while on a desert patrol in the Sinai. Above right: An M60A1 of the US 2nd Armoured Division fords a river on an exercise in America.

Right: An M60A2 lies in wait by a road during manoeuvres in Germany. The redesigned turret – capable of housing the 152mm gun/launcher – is clearly visible.

MERKAVA

The Merkava MBT, designed specifically to meet the requirements of the Israeli Armoured Corps, entered service in 1979. By modern standards it is a slow tank but it has excellent armour protection and can carry a large amount of ammunition for its 105mm L7/M68 type rifled tank gun which is also installed in Centurion, M48 and M60 MBTs of the Israeli Army. In designing the Merkava great emphasis has been placed on crew survivability as tank crews are expensive to train and, for a small country like Israel, always in short supply in time of war. The turret of the Merkava also has a relatively low cross-section to make it harder to detect and hit. A front wheel drive vehicle, the Merkava was first used in action in the Lebanon in 1982 (Operation Peace for Galilee) when it gave a good account of itself against the Syrian T72s, which were supplied by the Soviet Union.

The concept, design and construction of the Israeli Merkava main battle tank (MBT) was the result of a complex combination of factors stemming from an early recognition by Israeli defence planners of the need for self sufficiency in major weapon systems. During the 1948 war the Israelis suffered from an immense shortage of armour, relying on a few hastily repaired, near-derelict M4 Shermans procured by the Haganah, a handful of Cromwell and Hotchkiss tanks and some locally manufactured 'sandwich' armour-plated trucks. In 1949 Israel acquired a number of Shermans from scrap collections and ex-US Army dumps in Europe and these were modified for service with the Israeli Defence Force (IDF). During the 1950s most of Israel's armour was bought from France and the United Kingdom and included Shermans, AMX-13s and Centurions. Tank modification began to function in earnest in 1955 with the introduction of the Israeli-designated M50 Sherman, an up-gunned version of the M4A3 with a new engine. This was followed by the 'upgraded' Centurion, featuring a 105mm main gun and the Continental V12 engine, which was to see extensive combat experience in the 1973 Yom Kippur War.

In October 1966 two British Chieftain tanks arrived under a cloak of secrecy for extensive trials with the Israeli Army, with a view to future purchase. However, on the outbreak of the Six-Day War in June 1967 the British withdrew their offer to supply Israel with the new MBT; Israel's only remaining source of supply was the USA from whom they were buying M48s, M60s and M60A1s. The possibility of both Arab pressure on the US to curtail this vital supply and the fact that the Egyptians by the end of 1967 were amply restocked with T54s and T55s from the Soviet Union led the Israelis to realise the necessity for a

Previous page: With its bazooka side plates removed to avoid clogging, an IDF Merkava ploughs its way through heavy mud while on exercise on the Golan Heights. Top: The Merkava's low profile and uncluttered turret were key design features from the first experimental prototype (above) to its present-day active deployment with the IDF.

Far left: The Merkava's unique rear hatch provides both a rapid escape route from the combat compartment and the capacity for carrying battlefield casualties and infantry over fire-covered ground. Left: The 105mm gun, fitted with a thermal sleeve, is stowed for transport.

home-produced MBT if they were to maintain control over their military future.

Basic design work on a new MBT began in 1967, but it was not until governmental sanction for the project was gained in 1970 that work began in earnest on the detailed planning of the Merkava. General Israel Tal, veteran of the 1948 war, the 1956 Sinai campaign and the Six-Day War, headed the project. The variety of tanks deployed and the combat experience of the three conflicts were crucial in the design considerations for the new tank. Experience on the Golan Heights in the Six-Day War had proved that battlefield mobility was no substitute for armour protection, a lesson that was to be learned again to the great cost of tank crews in the early stages of the 1973 Yom Kippur War. As a consequence of Israel's small population and the time and finance needed to train men to the required standard, protection of the crew emerged as the most important design feature. Every part of the tank had to play its part in this respect; firepower came second and mobility third.

In comparison with the performance of other contemporary MBTs, the Merkava is slow and heavy, with a high combat weight of 56,000kg (123,480lb). Powered by a Teledyne Continental AVDS-1790-6A V-12 diesel engine developing 900hp at 2400rpm, it achieves a maximum speed of only 46km/h (29mph) compared with the American M1 and the German Leopard II's 72km/h (45mph). The original Merkava Mark 1, first delivered to the IDF in 1979, has coil-spring bogie-type suspension and Centurion road wheels but may be provided in the Mark 2 and Mark 3 developments with hydropneumatic suspension and possibly a 1200hp engine, increasing both its power and mobility on the battlefield.

While the Merkava lacks top performance capability, it is unique in its layout and emphasis on crew survivability. The turret and combat compartment are situated towards the rear of the vehicle, while the engine is placed in the vulnerable spot at the forward end. This design concept is found elsewhere only in the Swedish 'S' tank and provides the crew with considerable additional protection. The fighting com-

Below: In addition to the forward placement of the engine, crew protection is increased by the provision of detachable bazooka side plates.

Merkava Main Battle Tank

Crew 4
Dimensions Length (gun included) 8.63m (28ft 4½in); width 3.7m (12ft 1½in); height 2.64m (8ft 6½in)
Weight Combat loaded 56,000kg (123,480lb)
Engine Teledyne Continental AVDS-1790-6A V-12 diesel developing 900hp at 2400rpm

Performance Maximum road speed 46km/h (29mph); range (road) 500km (310 miles); vertical obstacle 1m (3ft 3in); trench 3m (9ft 10in); gradient 60 per cent; fording 1m (3ft 3in)

Armour Conventional cast and spaced
Armament One 105mm M64-L7 gun; one 7.62mm machine gun co-axial with main armament; one or two 7.62mm machine guns mounted on commander's and loader's hatches; one 60mm roof-mounted mortar; smoke dischargers

partment stretches the whole width of the vehicle, is air conditioned and comparatively spacious. Crew comfort is an important factor during prolonged engagements in climatic conditions such as those prevailing in the Sinai desert or Golan Heights. With the engine up front, the design provides for a rear door, making bailing-out a simple and protected manoeuvre which eliminates the hazards of a turret bail-out under fire or the trauma of a seized hull-bottom escape hatch in a blazing combat compartment. The rear door also allows other knocked-out tank crews to be taken aboard on the battlefield, and the tank can also be used as an infantry transport or command post with the main gun ammunition stowage rearranged.

The hull of the Merkava is constructed of cast and welded armour, manufactured after a long experimental and trial period at the Urdan factory near Netanya. The armour is spaced, providing the Merkava with increased protection against HEAT and HESH chemical action rounds, the spaced compart-

Above: Finished Merkava tanks in the factory grounds await delivery to their units having completed field tests.
Below: Flying black and green ceremonial pennants on its antennae, a Merkava is put through its paces in open country.

ments between the two layers of cast armour being filled with diesel fuel (not normally inflammable when hit). Additional armour protection is provided by side plates which help guard the tank's suspension.

Combat experience in the desert (where there is very little cover) emphasised the necessity for a tank design which offered the lowest possible target profile. The Merkava has a wide, forward-sloping armour-plated glacis with a cast armour front, and has a flat turret with a small cross section and long overhang at the rear. The low profile and main gun angles of plus 20 degrees elevation and minus 8.5 degrees depression allows the Merkava to remain virtually hidden while using its main gun in a hull-down position.

The main armament consists of a 105mm M64-L7 rifled tank gun manufactured by Israeli Military Industries, firing HEAT, HESH, APDS, APFSDS and phosphorous rounds. Secondary armament includes a 7.62mm co-axial machine gun and one or two 7.62mm machine guns mounted on the commander's and loader's hatches, a 60mm roof-mounted mortar and smoke dischargers. Much emphasis is placed on the training of tank gunners and commanders in the IDF and the Merkava is designed for quick firing main armament with a high probability of a first round hit.

The turret turntable is located on the hull floor plate, supported by rollers so that they rotate together, carrying the turret crew through 360 degrees either with the aid of the hydraulic power pack or through manual operation. The commander is provided with an access hatch and protective cover with his observation and aiming sights located close over the turret roof thereby allowing all-round observation with full protection. The Merkava fire-control system features an American Cadillac Gage stabilising system with four positions (two for traverse and two for eleva-

Above: An Israeli paratroop squad training with a Merkava tank crew. In combat conditions the tank-infantry team provides for mutual protection. Up to six foot-soldiers can be transported in safety, but when necessary they take up advance positions to protect the vehicle against potential tank-killers. Left: Operating the fire control computer in the gunner's position.

tion), a modified digital-control M13 ballistic computer and a laser rangefinder. The commander is provided with a panoramic 360 degree periscope with magnification ranging from x4 to x20, coupled or uncoupled to the aiming system. The Merkava's high vision instruments include the commander's passive night-sight, a white and infra-red driving light and a main searchlight. The tank driver command system consists of three main elements: a commander's operating handle, an audio command intercom and a visual display unit at the driver's station.

During the Yom Kippur War many Israeli tank crews ran out of ammunition during prolonged engagements and thus the Merkava has been designed to carry more main gun ammunition than most contemporary MBTs in the stowage area at the rear of the hull. The main-gun rounds are stored in heat-resistant containers, loaded by the rear access doors into side-bins. While 63 rounds is the standard load carried, the Merkava's full capacity is in excess of 85, almost double that of the Leopard 2 or the Soviet T72.

The Merkava first saw combat service during the 1982 invasion of the Lebanon, where it performed well against the T55s, T62s and T72s of the Syrian Army. Reports of future developments for the Merkava in the Mark 2 and Mark 3 variants suggest that, in addition to the installation of a more powerful engine and a superior suspension system, the Merkava may also exchange its combat-proven 105mm gun for a 120mm main armament. Without doubt the Merkava will remain the backbone of the Israeli armoured forces for many years to come.

Top: Specifically designed for desert warfare, the Merkava's wide sloping glacis and low hull-hugging main armament make it particularly suited to operating in the hull-down position. Above: A company of Merkavas advances over rough terrain to take up a defensive position in a wood on the Golan Heights.

S-TANK

Together with the British-supplied Centurion, the Bofors S-Tank forms the backbone of the Swedish Armoured Corps. This most unusual design has a three-man crew and is armed with a longer version of the standard 105mm L7 gun fed by an automatic loader. The gun is fixed to the hull and is aimed by traversing the whole tank and raising or lowering the hydropneumatic suspension. While this tank is ideal for Sweden's defence requirements, it does suffer from the major drawback that the 105mm gun cannot be aimed and fired on the move. In a confined space the lack of traverse could also be a major disadvantage as the S-Tank has to manoeuvre quicker than its rivals when there is a need to engage different targets.

When it first appeared in 1961, the Swedish Stridsvagn 103 MBT, or S-Tank, represented a revolutionary concept in tank design; not only because it was turretless, but because the entire vehicle was so engineered as to provide both the elevation and traverse for the fixed main armament.

In the late 1950s Sven Berge, head of tank design in the Swedish Army Ordnance Vehicle Division, set out to design a new armoured vehicle of radical design and low silhouette. Influenced by the French pre-World War II Char B and the more modern AMX13 which mounted its main armament in an oscillating turret, and exploiting experience gained in the design of self-propelled guns, Berge proceeded to design an assault gun – the IKV-103. This vehicle, which mounted a 105mm gun, was built by AB Landsverk and Bofors, and was later used as a test bed for the S-Tank concept to examine the problems of accurate alignment of the main gun. The IKV-103's system of two hydraulically-powered external levers which acted on the hull and individual tracks provided a means of precise control over track movement. Test results were encouraging and further trials were made using an American M4 Sherman medium tank chassis in which the hydraulic rams were placed inside the vehicle and the levers situated between the tracks and the hull.

In mid-1958 Bofors received the contract to develop the new tank. Although relatively inexperienced in vehicle design, the firm had amassed considerable knowledge in weapon systems design and, assisted by Volvo and Landsverk, proceeded with the development. In 1957 Landsverk had produced two prototypes for the KRV medium tank (which was discontinued) and their chassis were used as test beds for the hydro-pneumatic suspension and hydrostatic steering system of the forthcoming S-Tank. The first two prototypes were ready in 1961 and an order was placed with Bofors for 10 production models even before these had appeared.

The suspension system incorporated four Centurion road wheels on each side (without track return rollers) with a front drive sprocket and rear idler.

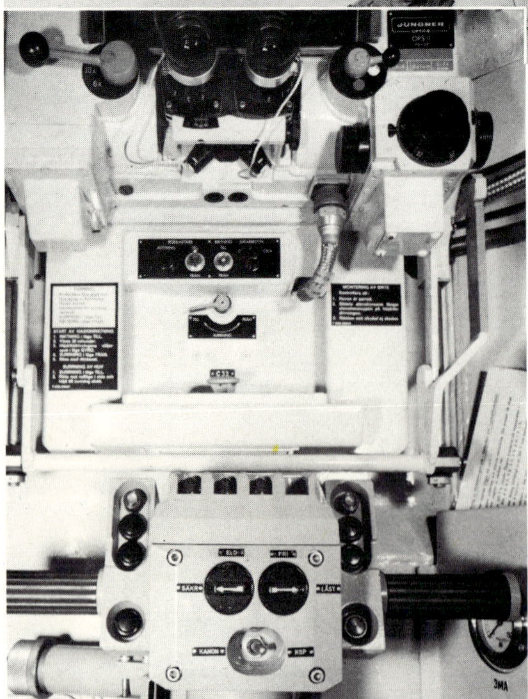

Top: S-Tanks advance across the battlefield supported by infantry and Swedish-built Pbv 302 APCs. Above: S-Tanks on the assembly line at the Bofors works. Since the late 19th century, Sweden has had an important arms industry. Left: The driver/gunner's position in the S-Tank showing the steering and suspension control unit with loading and firing buttons for the 105mm main armament and 7.62mm machine guns. Right: An S-Tank on winter manoeuvres puts its secondary armament to the test.

Previous page: An S-Tank accelerates up an incline on a forest training ground.

Rubber inserts on the dry pin tracks were added later. The road wheels are mounted on a system of hydraulically operated arms and by the elevation and depression of these arms the entire hull with its fixed gun may be raised or lowered to plus 12 degrees or minus 10 degrees while track length is compensated as it does so. Aiming in traverse is achieved by slewing the tank in its tracks. The concept is unusual but has proved perfectly satisfactory in operation through sound engineering and rugged components. Steering the tank and aligning the gun is controlled by a single unit fitted with handlebars, allowing the driver/gunner to operate the vehicle and engage targets with great speed. The commander is provided with a second control unit and can override the driver/gunner if necessary. The S-Tank is highly manoeuvrable and, because of its extremely low and flat silhouette, presents a very small target.

The third crew member, the radio operator, faces to the rear of the vehicle and is provided with controls which enable him to drive the tank backwards, a highly desirable characteristic in armoured warfare. He also increases the tank's visual protection by maintaining a rearward watch which gives the commander constant all-round observation.

The S-Tank is powered by the Rolls-Royce K60 engine which, like the Israeli Merkava's powerplant, is mounted forward in the hull and increases crew protection. A second 'booster' engine, the Boeing 533 gas turbine, is provided for negotiating rough terrain or for use in battle when rapid acceleration might be required. The gas turbine also aids cold-weather starting and provides an additional power source in case of diesel failure. Fuel tanks are mounted externally above the tracks.

Main armament consists of a rifled 105mm L74 gun which passes between the driver/gunner and the commander in the fighting compartment. It is a longer Bofors-manufactured version of the British 105mm L7 and is provided with an automatic loading device

Above: A pre-production prototype of the S-Tank with the original, domed, commander's cupola. Right and below: The S-Tank was the first tank to be fitted with a flotation screen permanently fixed to the hull. Once in position the vehicle can swim at 5.5km/h (3mph) propelled by its tracks.

Left: By dispensing with a turret, and mounting the main armament within the hull itself, the S-Tank's designers achieved a far lower silhouette for the vehicle than that of the conventional turreted tank. Compared with the Centurion on the right, the S-Tank offers a very limited target area and is ideally suited to operating in the hull down position.

Left and bottom: Two views of the S-Tank showing the elevation and depression of the fixed main armament. Pitch of the hull is precision-regulated by the vehicle's hydro-pneumatic suspension system. This innovation, however, does not allow the vehicle to fire on the move.

which feeds the 50 rounds available by hydraulic power from a magazine into the breech. The magazine is in two halves and various types of rounds can be employed including APDS, HE, HESH and smoke. Spare ammunition is stored at the rear of the vehicle. The efficiency of the automatic loader provides a rate of fire of between 10 and 15 rounds per minute which is substantially higher than most manually-loaded conventional tank armament. At the moment of firing additional stability is provided by the locking of suspension and tracks. The empty shell cases are automatically ejected from the rear of the vehicle.

Secondary armament in the early versions consisted of four Ksp-58 7.62mm machine guns box-mounted in pairs on the glacis and firing along the axis of the main armament. For a time the box to the driver's right was fitted with a 0.5in ranging gun, but this was later deleted. A further 7.62mm machine gun, mounted on the commander's cupola, was also provided. Eight smoke dischargers are fitted and some S-Tanks are fitted with Bofors Lyran flare launchers which can provide night-time battlefield illumination.

The S-Tank offers a high level of protection to its crew. In addition to the very flat profile of the glacis plate, ribbing on it decreases the possibility of rounds penetrating. As the pre-production models were evaluated further improvements were incorporated including a clamp for the protruding barrel of the main gun and a fume extractor fitted to the barrel, as on the Centurions in Swedish service. In addition two track return rollers were fitted on each side.

In the field of optics, the driver/gunner and commander are provided with a Jungner OPS-1 periscope/sight with wide angle vision and up to x18 magnification. With his cupola gyro-stabilised laterally and his OPS-1 in elevation, the commander enjoys a consistently high level of vision, assisted by four periscopes. The driver/gunner is provided with one periscope and the radio operator with two periscopes with armoured visors which can be lowered from inside the tank. A laser rangefinder has been developed, and this has led to an improvement in accuracy. Infra-red driving lights are standard though the vehicle has no infra-red searchlight.

Since deep-water obstacles abound in Sweden, the S-Tank is equipped with a flotation screen which is normally stowed flat around the tank's upper decking and can be erected in about 15 minutes to enable the tank to swim, powered by its tracks. This flotation gear was introduced as standard on the Stridsvagn 103B and retrofitted to the earlier A-series models. The vehicle is also fitted with a bulldozer blade which can be folded beneath the glacis plate when not required and manually brought into position for the excavation of a fire position. Stowage bins are provided at the rear of the tank.

The Stridsvagn 103A was formally adopted by the Swedish Army in July 1964 and the first of the 300

Bottom: A view of the S-Tank which clearly shows the bulldozer blade stowed beneath the hull front. The blade can be manually lowered and locked into position by the crew in five minutes and is used for digging out a firing position (inset) or clearing heavy obstacles. The depth of the cut made by the blade is controlled by altering the pitch of the hull.

vehicles now in service began to equip units in 1966. Production continued until 1971, and currently the tank equips six of the nine Swedish armoured brigades while the remaining three have the Centurion. The S-Tank has not been employed outside Sweden, but the United States, Great Britain and West Germany have all evaluated it. The tank is extremely cost-effective and has proved reliable in service, a tribute to both advanced design and sound Swedish engineering. At present 275hp Detroit Diesel 6V 53T engines are being installed which will prolong the tank's life until its planned successor appears in the late 1980s. This new AFV is likely to owe much to the S-Tank and is being developed jointly by Bofors and Hägland and Söner.

Although to date it has not been proved in battle, there is no doubt that the S-Tank provides the Swedish Army with an effective and well-armed vehicle which offers its crew good protection and will remain the mainstay of the Swedish Armoured Corps until the new MBT appears.

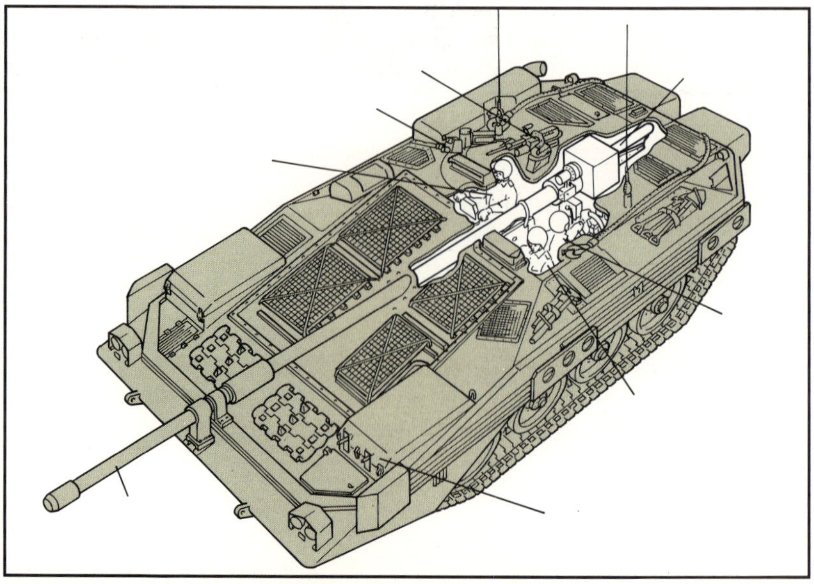

Stridsvagn 103B MBT

Crew 3
Dimensions Length (gun included) 8.9m (29ft 2½in); width 3.4m (11ft 2in); height (including machine gun) 2.5m (8ft 2½in)
Weight Combat loaded 39,000kg (85,979lb)
Engines One Rolls-Royce K60 water-cooled multi-fuel engine developing 240hp at 3650rpm; one Boeing 533 gas turbine developing 490shp at 38,000rpm

Performance Maximum road speed 50km/h (31mph); range 390km (245 miles); vertical obstacle 0.9m (2ft 11in); trench 2.3m (7ft 6in); gradient 60 per cent; water speed 6km/h (4mph)

Armour Details classified
Armament One 105mm L74 automatic gun; two 7.62mm machine guns mounted on the left hull front; one 7.62mm machine gun mounted on the commander's cupola; two four-barrelled smoke dischargers

Above: A rear view of the S-Tank. Ammunition is stored for safety at the rear of the vehicle; the S-Tank and the Israeli Merkava are among the few MBTs that include this feature. The S-Tank is also provided with additional stowage bins mounted on the hull rear. Left: The addition of ribbed armour in the form of a series of rectangular horizontal bars (here covered by wire grilles) to the well-shaped glacis plate provides the S-Tank with considerably increased protection against high-velocity armour-piercing rounds.

T54/T55 & T62

The Soviet T54/T55 and T62 series of MBTs have been produced in larger numbers than any other tanks in history. Although well over 10,000 of the more recent T64 and T72 tanks have been produced in the Soviet Union in recent years, the older tanks still remain in front-line service in large numbers. While the T54/T55 and T62 series have a number of major drawbacks when compared to Western tanks, such as poor fire control equipment and limited elevation and depression of the main armament, they have proved relatively simple to maintain and operate. Some countries have improved the T54/T55's night vision equipment, while Israel has fitted a 105mm rifled tank gun to the T54/T55 series. China also has a version of the T54, which is called the Type 59.

The Soviet Army's aggressive use of tanks during World War II laid the foundation for contemporary Soviet tactical thinking regarding the deployment of armour. Three of the present-day Soviet Tank Arm main battle tanks (MBTs) can trace their origins directly back to this period in the design of the T44, which was in turn a development from one of the most successful tanks ever built, the T34. The design of the T44, however, left much to be desired. Despite the various inadequacies of the turret design and the cramping of the crew, the main problem was still the installation of the 100mm D-10T main armament. By 1947, however, this had been ironed out and the result was the T54, a line of tanks destined to be produced in greater numbers than any other. Improvements on the T54 series led to the T55 and from there to the T62, all of which will be in Warsaw Pact service well into the 1990s.

The T54 has a welded hull divided into three compartments and carries a four man crew. The driver, commander and gunner are located on the left side of the vehicle while the loader occupies the right side of the T54's distinctive curved turret. The T54 is powered by a 12-cylinder water-cooled V-54 diesel engine (developing 520hp at 2000rpm) which incorporates a high percentage of alloy, thereby increasing the risk of fire to which the vehicle is particularly susceptible. The T54 is tiring to drive since no hydraulic assistance is provided, the torsion bar suspension is bouncy, and the dead track tends to shed itself during violent manoeuvres. Crew space is extremely limited, which contributes to fatigue in combat, for the T54 is a compact vehicle offering a low silhouette and a small frontal target area.

Main armament on the T54 is the 100mm D-10T, which began life as a naval gun and in early models was not stabilised. Although successive improvements have included stabilisation, the conditions under which the crew has to operate the gun and the fire control system are still far from ideal. While the one-piece cast turret rotates, the floor of the fighting compartment does not, which increases the risk of being crushed by the gun's breech block as the turret swings. Loading the long and heavy fixed-case ammunition is difficult and the loader has only three ready reloads available, the remainder being stored in the hull. The gun has to be elevated after each round has been fired for the case to be ejected, which greatly slows the firing sequence.

In comparison with the computer-assisted fire control systems of Western MBTs, the T54's fire control system is slow and inaccurate. Having identified the target through the TPK-1 binocular sight, the commander swings the turret into rough alignment and the gunner acquires the target through his TSh-22 telescopic sight. Ranging is accomplished using the commander's stadiametric rangefinder, after which ammunition type is selected and the target is engaged. Available ammunition on the T54 includes AP-T (armour piercing tracer), HE (high explosive), HE-Fragmentation, HEAT (high explosive anti-tank) and HVAPDS-T (high velocity armour piercing discarding sabot tracer). Secondary armament originally consisted of a bow-mounted 7.62mm machine gun (which was later deleted) and a 12.7mm DShK machine gun, for use against aircraft, mounted on the loader's cupola.

In 1955 a variant appeared and was designated the T54A. It had a D-10TG gun stabilised in the vertical plane and two years later it was followed by the T54B with its D-10TS gun stabilised laterally. Further improvements included infra-red sights for the commander and gunner and snorkelling equipment, which enabled the tank to ford rivers to a depth of over

Previous page: Soviet T55 crews parade in front of their vehicles while taking part in an armoured vehicle exercise. Above: A T55 travels at speed during Warsaw Pact manoeuvres in Czechoslovakia. The final production version of the T54 series, the T55, remained in production in the Soviet Union until 1981, some time after the T62 had been phased out of production. The T54/T55 was much cheaper to produce than the T62 and remains in service with some 45 countries all over the world.

4m (13 feet), an invaluable aid in an offensive.

In 1961 the newly designated T55 appeared which incorporated still further improvements on the T54 model. The T55 was fitted with the V-55, a more powerful engine which increased the power output to 580hp at 2000rpm, and a rotating turret floor was added. Externally the T55 is virtually the same as the T54, with the exception of the removal of a circular ventilator in front of the loader's cupola. Further improvements to the design were introduced in the T55A in 1963. These included a plastic and lead anti-radiation lining and an NBC (nuclear, biological and chemical) warfare system which automatically closes all hatches and shuts the engine down to warn the crew of NBC detection. Since sealing is incomplete, the crew still has to operate in NBC suits. The T54's hull machine gun was removed to provide stowage for an additional six rounds of main gun ammunition, as was the 12.7mm machine gun, although this has subsequently been refitted to many tanks. A laser rangefinder is now fitted over the main armament and infra-red equipment includes the L-2G

Above: The Israeli Army has always been ready to utilise captured weapons, including this T55 re-equipped with a new engine and a 105mm L7A1 gun. Below: A forward view of a Syrian T54A, knocked out on the Golan Heights during the Six-Day War of 1967.

Above: A Soviet tank commander looks through the turret periscope on his T54 tank. While adequate in most circumstances, the optics on Soviet tanks are not up to the standard found on Western equivalents.

searchlight, the OU-3GK sight for the commander, and periscopes for the driver, gunner and loader. This equipment has also been retrofitted into earlier T54s, as has the NBC fit.

Variants to the T54/T55 series include an ARV, bridgelayers and mine clearance vehicles, while the chassis provides the basis for the ZSU-57-2 AA gun and the SU-130 assault gun.

Since 1945 the T54 and T55 have seen considerable combat, notably in the hands of the Arab armies in the Middle East wars of 1967 and 1973. Some 400 tanks were captured by the Israelis and these have been modified by the installation of the Centurion L7A1 gun. In addition to Warsaw Pact countries, many Third World nations have been equipped with them and they have seen action in Vietnam, Iraq,

T54/T55 Main Battle Tank

Crew 4
Dimensions Length (gun included) 9m (29ft 6in); width 3.27m (10ft 9in); height (without AA MG) 2.4m (7ft 10in)
Weight Combat loaded 36,000kg (79,366lb)
Engine V-54 (V-55), V-12 water-cooled diesel developing 520hp at 2000rpm

Performance Maximum road speed 48km/h (30mph); range (road) 400km (249 miles); vertical obstacle 0.8m (2ft 8in); trench 2.74m (8ft 10in); gradient 60 per cent; fording 1.4m (4ft 7in). 4.5m (14ft 9in) with snorkel

Armour Conventional, 20-203mm (¾-8in)
Armament One 100mm D-10T gun; one 7.62mm machine gun co-axial with main armament; one 7.62mm machine gun in bow of tank; one 12.7mm DShKM anti-aircraft machine gun on loader's cupola

Top: The interior of the T62 showing the racked rounds of AP ammunition (left) and the large breech-block of the 115mm main armament (right). Above: Two views of the T62 turret, with the cartridge ejector (left). Left: Manoeuvrability has always been rated as an important factor in the design of Soviet tanks, and is demonstrated by this example crashing through thick mud.

Egypt, Sudan, Angola, Uganda, the Ogaden, India and Pakistan. The Soviet Union still deploys them in the Naval Infantry, where they are standard equipment, and in some Tank and Motor Rifle Divisions.

Soviet tank development has followed a linear progression and the T62 MBT, which appeared in 1965, is a further improved version of the T55. The T62 has an all-welded hull and a cast turret housing the 115mm U-5TS gun with a co-axially mounted 7.62mm PKT machine gun. As in the T55, three of the four crew members are aligned on the left side of the vehicle and crew conditions show little improvement.

The T62's main armament of the smooth-bore U-5TS gun represents a considerable improvement over the D-10 series. Gun depression has not been improved beyond the minus four degrees of the T54/T55, which severely limits hull-down operation, but the ammunition provided has greatly improved its fighting capability. APFSDS will reach out to 1600m (1750 yards) at a muzzle velocity of 1680m/sec (5510ft/sec) and will penetrate 300mm (11.8in) of armour at 1000m (1090 yards). HEAT and HE rounds are also carried by the T62 but the rate of fire still remains at about four rounds per minute. With this new gun and rangefinder the T62 can engage targets at longer ranges than the T54 or T55 but, despite its stabilisation system, the tank must stop before opening fire with its main armament.

The T62 is powered by a 12-cylinder V-2-62 diesel engine developing 580hp at 2000rpm. A compressed-air start system is fitted to augment the battery and pre-heating is available. Driving the tank remains a tiring business. Heat build-up in the driver's compartment cannot be alleviated by opening the hatch since when it is open the turret will not revolve.

Developments of the T62 include the T62A, which mounts a 12.7mm machine gun by the modified

Below: A T62 fords a water obstacle with its distinctive snorkel in position. Few rivers in central Europe would allow such an easy exit from the water as shown in this photograph, however. The T62 can snorkel for a maximum distance of 1000m (1100yds) at a top speed of 2km/h (1.25mph).

loader's hatch, and the T62K command tank with additional communications 'fit'. The T62K incorporates the T72's track and drive sprocket and there is also an ARV variant, the T62T.

Estimates of production figures for the T62 vary but it is likely that over 40,000 have been built in the Soviet Union. It has seen action in the Middle East and along the Sino-Soviet border. Foreign recipients include Afghanistan, Angola, Algeria, Cuba, Czechoslovakia, Ethiopia, India, Iraq, Libya, Mozambique, Somalia, Syria, Vietnam, the People's Democratic Republic of South Yemen and Israel (as war booty).

Compared to contemporary Western MBTs such as the Leopard or Abrams, the T62 is an unsophisticated machine. It does, however, have the advantage of being cheap and easier to manufacture, and in the context of Soviet numerical superiority and massed tank assault doctrine the T54, T55 and T62 tanks have shown that the Soviet Union is one of the most effective of armoured vehicle manufacturers.

T62 Main Battle Tank

Crew 4
Dimensions Length (gun included) 9.33m (30ft 7½in); width 3.3m (10ft 10in); height 2.39m (7ft 10in)
Weight Combat loaded 40,000kg (88,185lb)
Engine V-2-62, V-12 water-cooled diesel developing 580hp at 2000rpm

Performance Maximum road speed 46km/h (29mph); range (road) 450km (280 miles); vertical obstacle 0.8m (2ft 8in); trench 2.85m (9ft 4in); gradient 60 per cent; fording 1.4m (4ft 7in), 5.5m (18ft) with snorkel

Armour Conventional 20-170mm (¾-6¾in)
Armament One 115mm U-5TS gun; one 7.62mm machine gun co-axial with main armament; the 62A retains the 12.7mm DShKM anti-aircraft machine gun on the loader's cupola.

Above: A Czech T54, equipped with a short snorkel tube, enters the water on a river crossing in Bohemia.

Below: T62As of the Afghan Army trundle through the streets of Kabul on a parade designed to show the population the strength and resolve of the government. The T62A differs from the standard T62 only in having a 12.7mm anti-aircraft machine gun.

T64, T72 & T80

The Soviets still believe that the tank has a key role to play on the battlefield and since World War II have committed a large amount of the army budget to the production of new MBTs. The T64 was built exclusively for the Soviet Army whereas the T72 has been exported on a large scale and also made in Czechoslovakia, Poland and Yugoslavia. India is also expected to begin producing the T72 in the near future. Both the T64 and T72 feature a three-man crew and a 125mm gun fed by an automatic loader. Late production models of the T72 are called the T74 by the Soviets although the United States have designated this tank the T80.

Soviet defence priorities enable the USSR to produce a new-generation main battle tank (MBT) roughly twice as often as its Nato counterparts. In effect this means that a new Soviet MBT will be issued to the various Warsaw Pact armies when its successor is already on the drawing board, and will only last a decade before that successor is brought into service. Nato countries have tended to introduce a new tank design every 20 years or so and to make do by modifying and retrofitting their existing tank forces to keep abreast of up-to-date tank technology. On the surface this situation seems to be weighted in favour of Russian tank designers and Soviet MBTs might be expected to have benefitted from this less-interrupted flow of development and be more than equal to their Western counterparts. In practice, however, this is far from the case. Indeed, Soviet tank technology in many areas seems to be a half step behind.

In 1961 the T62 MBT entered production and eventually became exportable to Soviet allies and clients. In combat conditions during the 1973 Arab-Israeli War it proved markedly inferior to Israeli-crewed Centurions, its rate of fire, main armament, angle of gun-barrel depression and fire control system being particularly inefficient in comparison. Admittedly the Israeli Centurions had been extensively, and often brilliantly, modified in Israel but there was no disguising the fact that they were comparatively antique, having first been produced in 1945. The Soviet Union, however, could afford to view these failings with a certain degree of equanimity since their next generation tanks, the T64 and T72, were already in production, influenced not only by recent Soviet technical advances, but also by the known capabilities of the new generation of Nato MBTs, the German Leopard 1, the British Chieftain and the American M60. As the T64 and T72 entered service in quantity in the mid 1970s, the Russians could draw some satisfaction from the alarm voiced in Nato military establishments that they had produced two new, threateningly superior weapon systems.

The T64 and T72 are, to some extent, twins; both were developed from the same prototype and are in many ways very similar. The T64 was the first into production in the late 1960s but was soon followed in 1972 by the T72. After that, production of the T64 was much more limited – it was built at only one plant while five factories manufacture the T72. Although the T72 seems to be the favoured MBT, having entered service throughout the Warsaw Pact, the US Army considers the T64 to be its combat equal and, as it equips the important group of Soviet forces in East Germany, the Soviet Union obviously shares this belief.

A number of new ideas have gone into the design of the T64 and T72, of which the most striking is that the tanks are crewed by only three men. While the driver, commander and gunner remain, the human loader has been replaced by an automatic loading mechanism, potentially rectifying the T62's poor rate of fire. The introduction of an automatic loading system, however, raises a number of other problems since there is almost certainly a difficulty in making sudden changes in selection of the type of round required, and reports suggest that on occasion one of the gunner's limbs has been selected by the system's mechanical arm and loaded into the breech. In the West, doubts remain as to the reliability of the automatic loader and there has certainly been no move in Nato countries to introduce a three-man crew vehicle. The only non-Soviet MBT to feature this type of configuration is the turretless Swedish S-Tank.

Previous page: A column of Polish T72s rolls across open country. Unlike its sister tank the T64 which is exclusive to Soviet forces, the T72 presently equips all Warsaw Pact armies and has also been supplied to Algeria, Bulgaria, Cuba, Czechoslovakia, East Germany, Hungary, India, Iraq, Libya, Poland, Romania, Syria and Yugoslavia.

Above: The T62, the forerunner of the T72, on manoeuvres in the Soviet Union. Ten years after it appeared, the T62 was superseded by the T64 and T72 series but remains in service with a great many Soviet foreign clients, and saw considerable action in the Middle East in the 1973 Yom Kippur War.

A further innovation in the T64 and T72 design was the upgunning of the main armament. Prototypes and early production models of the T64 were fitted with the same smooth-bore 115mm gun as the T62, but the decision was soon made to change to the more powerful 125mm armament. As a result, all T64s were refitted with the heavier gun and T72s were similarly equipped from the outset. The 125mm weapon is formidably powerful and completely stabilised. Once the gunner has laid it on target, it will remain aimed at the same spot, theoretically providing a very high degree of first-round accuracy while on the move over rough country. In addition the T64 and T72 are fitted with new-technology fire control systems. The gunner is provided with a laser rangefinder, considerably more accurate than the stadiametric rangefinder on the T62, and an improved panoramic day-night sight. Apart from these initial improvements, the level of sophistication of the fire-control system has been upgraded whenever the pace of technological change dictates; in the 1981 Moscow Parade, T72s lacking the right-side aperture forward of the commander's cupola on the turret roof were spotted, indicating further changes to the system. Syrian experience with the 125mm gun, however, suggests that while this gun, with its very high velocity, is capable of blasting the turret right off an enemy tank, its accuracy is severely impaired by its lack of rifling. A complete

Above: A group of T64s awaits the order to move out while the crews confer over the map on Warsaw Pact manoeuvres.

Below: Crowds welcome a formation of T64s as they move proudly through the streets of an East German town.

МИР
СОВЕТСКАЯ ПРОГРАММА МИРА
XXVI
КПСС
НАШ КУРС: МИРНОЕ СОЗИДАНИЕ!
ДА ЗДРАВСТВУЕТ МИРОЛЮБИВАЯ ЛЕНИНСКАЯ ВНЕШНЯЯ ПОЛИТИКА СССР!

472

Far left: Tank crews prepare their T72s for a submerged river crossing. Snorkels are fitted while hatches and gun barrels are sealed. Left: Soviet T72 commanders salute as their vehicles parade majestically through Moscow's Red Square. Below left: A column of T72s, with Hind helicopter gunships in support, simulates a full-scale armoured assault. Below: A Soviet T72 crew makes a routine track inspection during Warsaw Pact manoeuvres. Bottom: With main gun and anti-aircraft machine gun at the ready, the crew of a dug-in-Iraqi T72 scans the horizon for any signs of hostile Iranian activity.

picture of the weapon's capability has yet to emerge, but during the first Israeli tank encounter with Syrian T72s in the Lebanon in 1982, 10 Syrian tanks were knocked out by the considerably lighter Israeli 105mm armament.

To complement the increase in firepower, the T64 and T72 are thought by the US Army to have much tougher armour protection than their predecessors. From its first appearance it seemed likely that the armour on the new generation was of the conventional cast and rolled type, rather than the British-developed Chobham type discernible by its characteristic flat-shaped turret front and sides. Since the turrets of the T64 and T72 are obviously of a shape not compatible with Chobham armour, some puzzlement greeted a 1978 US Army report that the new Soviet MBTs were protected by a new advanced type of armour, providing the same degree of protection as Nato's Chobham. The armour on the turrets of the T64 and T72 appeared to be no thicker than that on the T62, so it seemed unlikely that it was spaced, and it was thus concluded that the new armour was a laminate type, incorporating the latest developments in Soviet metallurgy.

Differences between the T64 and T72 have always been small. Essentially the T72 is a slightly larger, wider and heavier tank, fitted with a more powerful engine. Tracks on the T72 are wider, providing a lower ground pressure and a higher power-to-weight ratio, which makes it marginally more agile than the T64. On the T72 the searchlight was moved across the turret to improve the gunner's field of vision, while the 12.7mm turret-roof machine gun was given a pintle mounting to increase its angle of elevation. The roadwheel design is also different, the T64 having six small stamped wheels on either side whereas the T72 is fitted with six larger die-cast rubber-coated roadwheels. Both tanks mount a 7.62mm machine gun co-axial with the main armament, and are provided with snorkels for fording and a full NBC (nuclear, biological, chemical) fit; both can accommodate long-range fuel tanks and front-mounted dozer blades. Both vehicles have a command variant, T64K and T72K, with radio and navigation equipment similar to the T62K command vehicle and a 10m (32ft 9in) mast antenna supported by stays. A major drawback, however, is that neither command tank can travel with the mast erected.

By the early 1980s the T72 seemed to have been favoured as the main Soviet and Warsaw Pact army tank to stay in production and be updated when necessary, until it is replaced by the controversial Western-designated 'T80' about which little hard fact is known. In Western military establishments, no agreement has yet been reached as to whether it is a completely new Soviet MBT or merely an update on the T72. In March 1983 the US Department of Defense designated the new tank as T80 although Soviet sources refer to it as the T74. A great deal of fairly wild speculation has gone into the characteristics of the new tank and, while some consider it to be protected by a Chobham-type armour, photographs purporting to be of the T80 show it to be very similar to the T64/T72 series, only with modifications to the turret shape. The US Department of Defense, always anxious to improve its tank-design defence-budget allocation, supports the view that the T80 is a very real and potent threat to Western security and rumours abound that there may be an even more advanced tank currently under development in the Soviet Union.

Certain features of the T80 do, however, seem likely in the light of Soviet tank-development priorities. It is thought to retain the 125mm main armament with automatic loading system, with possible improvement to ammunition, and to introduce a digital fire-control computer and passive night-vision gear. It is also believed that the Soviets have abandoned the torsion-bar suspension fitted to the T64 and T72 in favour of a variable hydro-pneumatic type for improved performance. Such modifications have yet to be seen to be proved, however, and whether the T80 will prove a match for the capabilities of its latest Nato counterparts is in itself a matter of total speculation.

T72(T64) MBT

Crew 3
Dimensions Length (gun included) 9.24m (30ft) (**T64**: 9.1m – 29ft 8in); width (including skirts) 4.75m (15ft 5in) (**T64**: 4.64m – 15ft 1in); height 2.37m (7ft 8in) (**T64**: 2.3m – 7ft 6in)
Weight 41,000kg (90,388lb) (**T64**: 38,000kg – 83,775lb)
Engine Diesel developing 780hp (**T64**: Diesel developing 700hp)

Performance Maximum road speed 80km/h (50mph) (**T64**: 70km/h – 43mph); range (internal fuel tanks) 500km (311 miles); range (with long-range fuel tanks) 700km (435 miles); vertical obstacle 0.915m (3ft); trench 3.07m (9ft 9in) (**T64**: 2.72m – 8ft 10in); gradient 60 per cent

Armour Classified
Armament One 125mm gun; one 7.62mm machine gun co-axial with main armament; one 12.7mm anti-aircraft machine gun

Below: Although the T64's 12.7mm anti-aircraft machine gun can be fired remotely while the vehicle is closed down, the T72 featured a pintle-mounted machine gun with better elevation. Although more effective, this system exposes the commander to enemy fire while the machine gun is in use. Below centre: The various types of main-gun ammunition available on the T72 are displayed at the front of a tank's glacis. Ammunition includes HEAT, APFSDS and two types of HE fragmentation round. Bottom: Like all Soviet MBTs, the T72 has the facility to carry long-range fuel tanks at the rear of the vehicle.

Above: One of the early photographs of the Soviet T74 MBT which confirm the view that the T74 is merely an update of the T72 and not the awesome opponent that the US Department of Defense would have the world believe. Right: A US 'artist's impression' of the T80 which shows the tank to have a Chobham type of armour.

AMX13

The AMX13 has been one of the most successful of all French armoured vehicles developed since World War II. The first production vehicles were completed at the Atelier de Construction Roanne in 1953, and in 1985 the tank was still in production. Today it is built by Creusot-Loire at Chalon-sur-Saône for the export market only. Current production models are available with passive night vision equipment, NBC (nuclear, biological, chemical) system and laser range-finder, and the petrol engine can be replaced by a Detroit Diesel which not only gives increased operational range but reduces the fire risk. Using the same basic chassis a complete family of vehicles has been developed with total production amounting to 15,000 units.

The design, development and production of any new armoured fighting vehicle (AFV) is a lengthy and expensive process and one of the keys to a tank's long-term success is the vehicle's ability to provide a sound working platform for a number of other weapon systems and specialised battlefield equipment. Successful designs such as the US M48 series and the German Leopard I have provided the basis for a great many weapon types, from self-propelled and anti-aircraft guns to specialist engineering vehicles and flame throwers. The economic advantages of exploiting a tried and tested chassis in as many service roles as possible are enormous. In this respect the French AMX13 light tank is an excellent example: developed in the early 1950s, it has given birth to many variants and has seen service in some 25 countries.

Immediately after World War II, in line with government policy to re-equip with all-French vehicles and guns, the French Army's AFV design and development centre – the Atelier des Constructions d'Issy-les-Moulineaux (AMX) – was set to work on a programme for three new vehicles; these were the AMX50 heavy tank, the Panhard EBR-75 heavy armoured car and the AMX13 light tank. The original specification required the AMX13 to be airportable, although this was subsequently abandoned, and it was intended that the tank be primarily employed as a tank destroyer and reconnaissance vehicle. Work on the design went ahead quickly and by 1949 the first prototype was complete. Three years later the AMX13 went into production at the Atelier de Construction Roanne where it was produced until the early 1960s, when the Roanne works switched to construction of the AMX30 main battle tank (MBT) and the AMX13 was transferred to the Creusot-Loire factory at Châlon-sur-Saône.

One of the most interesting and unique features of the AMX13's design is its two-piece oscillating turret. The lower part of the turret is fitted to the turret ring in the normal way, while the upper part, which houses the main armament, is mounted on two trunnions allowing it to pivot and elevate and depress through 18 degrees. Since the whole top part of the turret moves up and down with the main armament fixed within it, this has allowed for the installation of a breech-aligned automatic loader in the turret bustle, and the gun is fed by two revolving magazines with a

AMX13 Light Tank

Crew 3
Dimensions Length (gun included) 6.36m (20ft 10in); width 2.5m (8ft 2in); height 2.3m (7ft 6in)
Weight Combat loaded 15,000kg (33,070lb)
Ground pressure 0.76kg/cm^2 (10.8lb/in^2)
Engine SOFAM 8Gxb eight-cylinder water-cooled petrol engine developing 250hp at 3200rpm

Performance Maximum road speed 60km/h (37mph); range (road) 350km (218 miles); vertical obstacle 0.65m (2ft 2in); trench 1.6m (5ft 3in); gradient 60 per cent; fording 0.6m (2ft)

Armour Conventional 40mm (1½in) maximum
Armament One 90mm gun; two 7.5mm or 7.62mm machine guns – one co-axial with the main armament, the other mounted on the commander's cupola; four smoke dischargers, two on each side of the turret

Previous page: An Israeli light tank of Colonel Amnon Reshef's AMX13 battalion rolls through Gaza on 6 June 1967, the second day of the Six-Day War. Reshef's AMX13s provided the crucial armoured support for Colonel Raphael Eitan's paratroop brigade in the hard-fought battle for control of the Gaza Strip. Top: An early prototype of the AMX13 ploughs through a sea of mud while on trial with the French Army. The early models were fitted with a 75mm main gun while the current standard AMX13 (above) mounts a 90mm fixed main armament in an FL-12 oscillating turret. Right: An Israeli 75mm AMX13 on reconnaissance enters a deserted town in the Sinai during a lull in the fighting in the Six-Day War.

full capacity of 12 rounds. Although this provides a very high rate of fire for the first twelve shots, it has a serious drawback on the battlefield in that the magazines have to be replenished from outside the vehicle. The first service type, the AMX13 Model 51, was fitted with the FL-10 turret mounting a 75mm gun, but in the mid-1950s a variant was produced for use in Algeria with a shortened 75mm armament housed in a slightly modified FL-11 turret. The current standard AMX13 is an upgunned 90mm version fitted with an FL-12 turret while the export model, the AMX13 Model 58, mounts a 105mm main gun. In the current 90mm model, 32 rounds of fin-stabilised HE (high-explosive), HEAT (high-explosive anti-tank), smoke and canister ammunition are carried and the main gun, with a muzzle velocity of 950mps (3117fps), can penetrate up to 320mm (12½in) of armour. Secondary armament can consist of two 7.5mm or 7.62mm machine guns, one mounted co-axially with the 90mm gun, and the other, which is optional, on the commander's cupola. Four smoke dischargers, two on either side of the turret, are also provided.

In line with modern developments in anti-tank weaponry, the AMX13 can also be fitted with ATGWs (anti-tank guided weapons), further enhancing its role as a tank destroyer. While retaining its 75mm gun, the anti-tank missile version of the AMX13 Model 51 mounts four SS-11 ATGWs in pairs on the turret front and can engage long-range targets up to an effective range of 3000m (3280yds). The French have also experimented with a model mounting six HOT *(haut-subsonique optiquement téléguidé)* ATGWs, although this was only taken to prototype and never adopted by the French Army.

Despite its offensive capabilities the AMX13 light

As new and more sophisticated anti-tank weaponry became available, the French experimented with two different types of missile, the HOT system (top) and the SS-11 (above centre). After trials the HOT-armed version was scrapped while the SS-11 AMX13 was adopted for service by the French Army. Above: An AMX13 armed with a 105mm main gun, a version specifically produced for the export market. Left: A knocked-out Israeli 105mm Mk 61 self-propelled howitzer which is based on the chassis of the AMX13 light tank. This model is no longer used by Israel.

Left: The AMX13-DCA anti-aircraft variant which mounts two 30mm cannon on a modified chassis. Below: The Dutch Army have further modified the AMX-VCI APC model to accommodate the American TOW anti-tank missile system. Bottom: A French Army AMX 105mm Mk F3 self-propelled howitzer. Two spades at the rear anchor the chassis while firing.

tank is a highly vulnerable vehicle. The hull, which is of all-welded steel construction to a maximum thickness of 40mm (1½in), although well sloped, provides little protection when up against high-velocity anti-tank rounds and guided missiles. To compensate for this, the AMX13's designers concentrated on keeping its size to a minimum and maximising its battlefield mobility. Along the same lines as the Soviet T64 and T72 MBTs, the introduction of an automatic loading system dispensed with the need for a fourth crew member, thereby reducing the size of the turret. Crew space is kept to a minimum and the maximum height of an AMX13 crewman is limited to 1.72m (5ft 8in). The crew consists of a driver seated in the front of the hull with the engine to his right, while the other two members, the commander and gunner, occupy the turret space to the left and right of the main armament. The AMX13 is powered by the Renault-built SOFAM 8Gxb petrol engine which, with the tank's low combat weight of 15,000kg (33,070lb) and its rugged torsion-bar suspension, gives it a high level of cross-country performance, even though its maximum road speed is only 60km/h (37mph). There are five road wheels on either side with the drive sprocket at the front and the idler at the rear; either two, three or four track return rollers are fitted.

Although the AMX13 remains in service with a number of countries' armed forces, it is by modern standards a relatively unsophisticated machine. The early production models had no night-vision equipment although most were subsequently retrofitted with infra-red driving lights and, in some cases, turret-mounted infra-red searchlights. Passive night-vision gear and a laser rangefinder are now available. The AMX13 light tank version also lacks NBC

Left: A battery of AMX 105mm Mk 61 SPGs on manoeuvres with the French Army. Below: The AMX-VCI armoured personnel carrier which utilised an extended version of the AMX13 chassis and also provided the basis for a number of variants. Firing ports are provided at the sides and rear of the vehicle and the turret mounts a 12.7mm machine gun.

(nuclear, biological and chemical) protection and no provision is made for amphibious use although it can ford water obstacles of up to 0.6m (2ft) in depth without preparation.

Apart from its light-tank role, the AMX13's robust chassis and powerful engine have also provided the basis for a great many other combat and support vehicles. The most extensive exploitation has been in the production of a group of vehicles centred around the AMX-VCI APC which features the same eight-cylinder SOFAM engine mounted in a slightly lengthened version of the AMX13 chassis. When the original AMX-VCI APC went into production in 1957 it carried ten infantrymen in addition to a three-man crew, but since then a number of different vehicles have been developed from this basic design. These include a version mounting a 20mm cannon turret, 81mm and 120mm AMX-VTT/PM mortar carriers, the VTT/PC command vehicle, the AMX-VCG pioneer vehicle, the VTT/TB unarmed ambulance and a battery command vehicle for use with artillery. Several missile-armed models have also been produced carrying ENTAC and SS-11 missiles, while an anti-aircraft version armed with two Roland surface-to-air missiles has been developed for trials. The Dutch Army has further modified the APC to accommodate the US TOW anti-tank system. Further variants for carrying cargo, supplies and artillery ammunition have also been manufactured.

Apart from the APC variants, the AMX13 has also provided the basis for two SPGs (self-propelled guns) and an anti-aircraft cannon system. The first of the SPG versions, the Mk 61 or Model A, appeared at the same time as the light tank and consists of an OB 105-61-AU 105mm howitzer mounted on a modified

AMX chassis. Weighing 16,000kg (35,274lb) loaded, it can fire a shell to a range of 15,000m (16,400yds), carries 56 rounds of ammunition and is served by a crew of five. The second SPG version, which came into service in the 1960s, is the 155mm Mk F3. This vehicle mounts the OB 155-50-BF gun and can fire the French Mk 56 and US M107 HE shells, the Brandt rocket-assisted projectile, as well as smoke and flare rounds. The Mk F3 is supported in combat by the AMX-VCI APC which carries the remaining six of the eight-man crew and further supplies of ammunition. It is also used in conjunction with the AMX-VCI battery command vehicle.

For anti-aircraft defence, the French Army is equipped with the AMX13-DCA which mounts two 30mm Hispano-Suiza HSS 831A cannon in a cast-armour turret on an AMX13 chassis. The guns can be fired at a cyclic rate of fire of some 600rpm from each barrel and have an effective range of 3000m (10,000ft). The AMX13-DCA also features the RD 515 Oeil Noir 1 (Black Eye) radar system which can detect targets to a range of 12km (7½ miles) at an altitude of 3000m (10,000ft).

The AMX13 series also includes two engineering vehicles, the *char de dépannage* Model 55 ARV (armoured recovery vehicle) and the *poseur de pont* or AMX13 bridgelayer. The ARV is fitted with two winches, an A-frame, various pieces of lighting equipment and tools for repair work. Armament consists of one 7.5mm or 7.62mm machine gun and smoke dischargers. The bridgelaying version mounts a scissors-type bridge on a modified AMX chassis and can span obstacles of up to 7.15m (23ft 3in).

It has been estimated that over 4500 AMX13 vehicles have been produced and while in the early 1980s it remained the standard light tank in service with the French Army, many of its variants have been replaced with more sophisticated equipment. Although the APC series was subsequently provided with NBC fit, it was replaced by the diesel-powered AMX-10P mechanised infantry combat vehicle family which is fully amphibious. The Mk 61 and Mk F3 SPG variants are being phased out in favour of the new French 155mm GCT guns which are based on the heavier AMX30 chassis and feature an automatic loader and fully-enclosed NBC protection.

The AMX13 series was very much a product of its generation, designed to meet the military requirements of the time. The light tank saw action with the Israeli armoured forces in both the 1956 Sinai campaign and the 1967 Middle East War but has since been retired, the Israeli command favouring the deployment of heavier and better protected MBTs. But in countries without the stringent up-to-date defence requirements of Israel, the AMX13 still has a role to play and is deployed in a number of Central and South American countries including Chile, Argentina, Peru, Ecuador and El Salvador, as well as with the armed forces of Indonesia, Morocco, Tunisia and Singapore.

Below: The AMX13 *char de dépannage* armoured recovery vehicle carries a crew of three and is fitted with a 15-tonne capacity winch and a five-tonne A-frame. Below centre: An AMX-VCG pioneer vehicle uses its A-frame and dozer blade to carry a bundle of logs, to drop into an anti-tank ditch. Bottom: An AMX13 *poseur de pont* swings its scissors bridge out over a river.

SCORPION

The Alvis Scorpion has been one of several successful armoured vehicles developed in Great Britain since World War II. Since the first production vehicles were completed by Alvis at their Coventry plant in 1972, over 3000 vehicles have been built for the home and export markets. In addition to being used by the British Army it is also in service with Belgium, Brunei, Honduras, Iran, Ireland, Kuwait, Malaysia, New Zealand, Nigeria, Oman, Tanzania, Thailand, the Philippines, the United Arab Emirates and the Royal Air Force Regiment. The basic chassis has been modified to accommodate a wide variety of vehicles to meet almost every battlefield requirement, ranging from reconnaissance to the carriage of supplies and the evacuation of wounded.

In the late 1950s the British Army issued a requirement for a vehicle called the Armoured Vehicle Reconnaissance (AVR). Eventually it was decided that no single vehicle could carry out all of the missions envisaged so two vehicles were developed. The Combat Vehicle Reconnaissance (Wheeled) became the Fox, and the Combat Vehicle Reconnaissance (Tracked) was to evolve into the Scorpion.

The CVR (T) had to have a fast top speed, be air-portable and be of a weight which would not exceed 8200kg (18,000lb), thereby making it lighter by a third than the Saladin, which it later superseded. But at the same time the new vehicle would have to maintain its independence from back-up units for fire support and communications. The most demanding specification was that, as a multi-role vehicle, it would be called on to perform three major duties: fire support, reconnaissance and anti-tank. Several experimental vehicles were put forward with main armaments ranging from 76mm to 105mm, but they were all far too heavy; and as the designers had drawn a blank in this field of research, it was decided to develop a basic vehicle from which a whole range could be developed by equipping them to meet the various roles demanded of them.

By 1963, the first outline of the new project, known as TV15000 Test Vehicle, was in hand. The results of TV15000 matured into a mobile test rig, which

incorporated most of the main engine, transmission and suspension elements which were to become standard on the final, Ministry of Defence-approved version of the CVR(T). In September 1967, satisfied with the initial trials, the MoD placed a contract order with Alvis of Coventry to supply 17 prototype vehicles. The new vehicle was to be called the FV101 Scorpion.

The first Scorpion prototype rolled out of the Alvis plant in January 1969, and a year later all of them had been delivered. There now followed exhaustive field trials in all extremes of terrain and climate. Following the satisfactory conclusion of the field trials, Alvis were awarded a production contract for 2000 Scorpion vehicles in May 1970. This order was soon increased when the Belgian Army placed its own order for 700 vehicles.

Above left: Armed with Swingfire missiles, with a machine gun mounted on the roof, an FV102 Striker prepares to fire. Above right: A forward view of the FV107 Scimitar; four such Scorpion variants served in the Falklands. The highly accurate 30mm Rarden gun carried by the Scimitar makes it very effective against soft-skinned vehicles. Thus the Scimitar would be capable of taking on the latest Soviet MICVs with success.

Two years later the first production model was completed. Over the next few years, Alvis were to produce no fewer than eight variants in the Scorpion range: the FV101 Scorpion (with a 76mm gun or a 90mm gun), FV102 Striker, FV103 Spartan, FV104 Samaritan, FV105 Sultan, FV106 Samson, FV107 Scimitar and finally Streaker (without FV prefix number). All of the Scorpion family of CVR(T)s share a number of common design features, such as the engine, tracks, suspension and running gear, which help in production and maintenance, and keep costs down.

All of the CVR(T)s are constructed mainly from a

Page 77: Two FV101 Scorpions advance across rough terrain in an armoured vehicle training area. Above left: This Scorpion is armed with the larger 90mm gun made by Cockerill in Belgium; the intention is to provide the Scorpion with greater battlefield punch.

Right: The FV103 Spartan, the armoured personnel carrier of the Scorpion range. The Spartan can also carry radar and night observation devices, and Blowpipe missiles.

Left: An FV107 Scimitar (distinguishable by its long 30mm gun) and a Scorpion advance together on an armoured patrol.

wheels, idler at the front and drive sprocket at the rear. The first and last road wheel stations have a hydraulic shock absorber.

The major feature of the PT-76 is its amphibious capability. Before entering the water the trim vane is erected at the front of the hull – to stop water rushing up the glacis plate and so pushing the vehicle further down into the water – and the two electric bilge pumps are switched on. When afloat the PT-76 is propelled by two water-jets mounted at the rear of the hull, one

Above: Mounted on a PT-76 chassis, a ZSU-23-4 anti-aircraft gun system. Other weapons based on the PT-76 chassis include the Frog-3 surface-to-surface missile (right) and the triple SA-6 surface-to-air missile (far right). Below: A line of PT-76s issues forth from a landing ship, trim vanes fixed.

on either side with the water entry port being located in the hull side above the fifth road wheel. When the driver wishes to go left the left water jet is covered and to go right the right one is covered. One of the drawbacks of this amphibious propulsion system is that the entry ports can easily be blocked by weeds or other debris. Maximum water speed is 10km/h (6mph) and the vehicle can be safely used in amphibious operations at sea as well as crossing inland waterways and lakes. One major problem, however, is actually leaving the water and shallow-angled exit points have to be chosen.

The fuel tank contains 250 litres (55 gallons) of fuel and is located in the right side of the engine compartment, but additional fuel tanks can be fitted over the rear engine decks to increase operational range. The PT-76 is not fitted with an NBC system and when originally built did not have any infra-red night vision equipment, but this has since been provided for the driver.

The PT-76 has formed the basis for a complete family of armoured vehicles, the first to enter service being the BTR-50P amphibious armoured personnel carrier. An improved version of this has been manufactured in Czechoslovakia under the designation of the OT-62. The Frog missile systems (marks 2, 3, 4 and 5) also use a slightly modified PT-76 chassis. Components of the PT-76, varying from just the track and roadwheels to almost the complete chassis, have been used in the ZSU-23-4 self-propelled anti-aircraft gun system, the SA-6 Gainful missile system, GSP heavy amphibious ferry and the PVA amphibious crawler tractor, to name but a few. Many of these vehicles have also been exported on a wide scale,

PT-76

Crew 3
Dimensions Length (gun included) 7.63m (25ft 1in); width 3.16m (10ft 4in); height 2.22m (7ft 2in)
Weight Combat loaded 14,000kg (30,865lb)
Ground pressure 0.48kg/cm^2 (6.83lb/in^2)
Engine V-6 water-cooled six-cylinder engine developing 240bhp at 1800rpm

Performance Maximum road speed 44km/h (27.5mph), 10km/h (6mph) in water; range (road) 260km (162 miles); vertical obstacle 1.1m (3ft 5in); trench 2.8m (9ft 3in); gradient 70 per cent; fording amphibious

Armour 10-15mm
Armament One 76.2mm D-56T gun; one 7.62mm machine gun mounted co-axially with the main armament

especially to the Middle East and Africa.

The scale of issue of the PT-76 within the Soviet Army has varied over time but in the early 1970s, for example, each Soviet motor rifle division and tank division had a total of 19 of these vehicles. The PT-76 is used mainly as a reconnaissance vehicle ahead of the main units but has also been used for crossing water obstacles in the first wave of the attack and for artillery support while a beach-head is being established. By early 1984 it had been replaced in front-line units by specialised reconnaissance versions of the BMP mechanised infantry combat vehicle, although it was still being used by Soviet marines. The PT-76 has been exported to a number of countries including Finland, India, Vietnam and Egypt.

China received a number of PT-76 light amphibious tanks from the Soviet Union and subsequently undertook production of an improved vehicle called the Type 63. This is slightly longer, wider and heavier than the PT-76 and retains its amphibious characteristics. Its most important feature is the new turret which is very similar in appearance to that of the T54/55 tank, although it is constructed of cast sections welded together and is armed with an 85mm gun.

The PT-76 was used in combat in Vietnam in the 1960s and when first encountered took the Americans and South Vietnamese somewhat by surprise, as they did not expect tanks to be employed by the North Vietnamese, although they were easily destroyed by air strikes and infantry anti-armour weapons. They have also been encountered by the South African Army during their clashes with SWAPO (South West African People's Organisation) in Angola where they were engaged and defeated by AML armoured cars armed with turret-mounted 90mm guns. The Indian Army deployed them in the invasion of East Pakistan (now Bangladesh) in 1971 where their amphibious capabilities proved highly effective. The Chinese version was probably used in the Chinese invasion of Vietnam in 1979. Although the PT-76 light amphibious tank is obsolete by modern standards it will doubtless remain in service for many years in those countries outside the Warsaw Pact, and its low ground pressure and amphibious capability make it a very useful vehicle in many tactical situations.

Below: A PT-76 of the Indian Army stands guard by the Intercontinental Hotel in Dacca at the conclusion of the war with Pakistan in 1971. The PT-76 proved its usefulness in the many amphibious crossings that were such a characteristic of the campaign in East Pakistan.

Above: The Chinese development of the PT-76 – the Type 63 – which features a new and enlarged turret mounting an 85mm gun. Although larger and heavier, the Type 63 retains the amphibious capabilities of the PT-76. Right: A PT-76 of the Finnish Army travels along a snowy track in support of infantry operations.

ARMOURED CARS

The armoured car and the smaller scout car still fulfil an important role in many armies around the world. In time of war their main job is to find out the position and strength of the enemy and report this back to higher command so that maximum use can be made of the main mechanised units. In peacetime armoured cars are often widely used for internal security operations where their speed and quietness are very useful. In many of the smaller armies, the armoured car is the single most important vehicle – some modern armoured cars have the firepower of a World War II tank.

The armoured car is not only the oldest type of fully automotive armoured fighting vehicle (AFV), but also the one which has changed least since it was conceived. At the end of World War I, such design features as good ground clearance and dual capacity front-and-rear steering were all established and since then the only important changes have stemmed from expanding technology which has provided improved performance and better armament. However, the purpose-built armoured car is a comparatively recent innovation, the first generation being tailored to fit existing commercial vehicle chassis; the first significant breakthrough was made in 1929 by Daimler-Benz with an experimental vehicle known as the ARW-MTW (Achtradwagen-Mannschaftstransportwagen, or eight-wheeled personnel carrier) in which each wheel was independently driven and suspended.

The Daimler scout car and the Daimler armoured car, which equipped the British armoured car regiments for several years after World War II, both owed a great deal to the ARW-MTW. Work on the former, a turretless 3048kg (6720lb) four-wheel drive vehicle with pre-selected transmission, began in 1937; nearly 7000 were built, being used for reconnaissance and liaison duties. The Daimler armoured car was a scaled-up version of the scout car weighing 7610kg (16,800lb) and mounted a turret armed with a 2-pounder gun and a co-axial machine gun; the 2-pounder armour-piercing performance could be improved by fitting a Littlejohn adaptor, which applied squeeze to the shot, so causing it to accelerate beyond its normal muzzle velocity.

During the period in which Great Britain granted independence to her former colonies, her armoured car troops were regularly employed on counter-insurgency tasks in various parts of the world, notably during the long but successful war against communist guerrillas in Malaya. In 1947 the War Office, as it then was, issued a requirement for the Daimler scout car's replacement; named the Ferret, it entered service in 1952 and was again manufactured by the Daimler

Previous page: A Ferret Mk 2/3 in Northern Ireland. Above: A Ferret Mk 2/6 fires a Vigilant ATGW. Right: The flotation screen raised on a Ferret Mk 4. Below left: A Saladin exercising on Salisbury Plain. Below right: A Saladin of 17/21st Lancers on patrol in Cyprus.

Company. In many respects the Ferret was very similar to its predecessor, but it was better protected, was larger, had a longer wheelbase and possessed a much superior cross-country performance. The Ferret Mark I, like the original Daimler scout car, was turretless, but the Mark II and subsequent Marks were fitted with a turret mounting a single machine gun, and in the early 1960s some Mark IIs were armed with four Vickers Vigilant ATGWs (anti-tank guided weapons), two in launching racks on either side of the turret and two stowed in the hull. The later Mark IV, weighing 5400kg (11,905lb) possessed a flotation screen and the final Mark V had a redesigned turret armed with four Swingfire ATGWs ready to fire, and two reloads. The Ferret is still operational with the British Army, and with many Commonwealth armies and some other countries. It has proved highly suitable, not only for its intended roles, but for peacekeeping missions as well.

The War Office also ordered the Daimler armoured car's replacement, the Saladin, in 1947 but Alvis, the manufacturers, did not go into quantity production until 1958. The reason for this was that the Malayan Emergency diverted those resources which had been earmarked for the Saladin into building the Saracen armoured personnel carrier, which employed the same six-wheel torsion bar suspension, and for which there was an urgent demand. The Saladin/Saracen suspension possessed steering on the leading four wheels and could simultaneously lose one wheel on each side to mine damage and still remain mobile. Armed with a powerful 76mm gun – from which the Scorpion light tank's main armament was developed – and a co-axial machine gun, the Saladin served with distinction throughout the fighting in the Radfan and the Aden Emergency, the Indonesian confrontation with Malaysia, and the Nigerian civil war. It remains one of the most successful armoured car designs ever built and has been purchased by several Commonwealth and foreign armies as well as by the West German Border Police. It weighs 11,590kg (25,550lb), is powered by a Rolls-Royce B80 engine producing 160 bhp, has a maximum speed of 72km/h (45mph), and is manned by a crew of three.

The most recent addition to the British armoured car fleet is the Fox scout car, which was intended to replace the Ferret in the 1970s, although in fact the latter remains in service. While the layout of the two vehicles is very similar, the Fox has a two-man turret armed with a 30mm cannon; one-man turret versions are armed with either a 25mm Hughes Chain Gun or the Milan (*missile d'infanterie léger anti-char*) ATGW system. Secondary armament consists of one 7.62mm machine gun. Like the Scorpion series, the Fox is constructed with aluminium armour and employs the same Jaguar 4·2 litre engine. It weighs 6386kg (14,080lb), has a maximum speed of 105km/h (65mph), and is capable of operating within a temperature variation of −25°C to +50°C (−13°F to +122°F). The vehicle was designed with air-portability in mind and three can be carried aboard a C-130 Hercules transport.

After World War II the US Army continued to use the six-wheeled M8, armed with a 37mm gun and co-axial machine gun, until the early 1950s; used by the French, it saw much action in Vietnam. The US then lost interest in armoured cars, their duties being carried out by light tanks or small armed but unarmoured wheeled vehicles with a good cross-country performance. However, in the early 1960s the Cadillac Gage Company developed its Commando Multi-Mission Vehicle, which could serve equally well as an armoured car or as a wheeled armoured personnel carrier. The Commando can be fitted with a variety of turrets and weapon systems to suit its user's requirements – reconnaissance, convoy escort, internal security, and so on. It is powered by a Chrysler 361 engine producing 200 bhp and has a top

Top: The Fox armoured car suffers from a high centre of gravity and drivers require special training. Centre: A Fox adapted to carry the Milan ATGW. Above: The US M8 armoured car, made by the Ford Motor Company.

speed of 88km/h (55mph); it can also swim without preparation, being propelled by the tread of its own tyres when afloat. In addition to serving with the US Army, the Commando has proved popular with the armies of a number of Third World countries.

By the end of 1941 the Soviet Union's fleet of BA armoured cars had virtually ceased to exist and it was not until almost 20 years later that the next Russian design appeared. It was the amphibious BRDM-1 containing a number of interesting features, including two pairs of retractable wheels in the centre of the chassis which can be engaged on difficult going, a central tyre-pressure control system, and a single hydrojet propulsion unit for use when afloat. Several types have been developed, the standard model mounting a 7·62mm machine gun, others with mountings for Snapper, Swatter and Sagger ATGWs. The BRDM-1 has a maximum speed of 80km/h (50mph) and a crew of between two and five, depending on type. In the early 1960s a replacement for the BRDM-1 entered service, the BRDM-2. This has a different layout, more powerful engine and a new one-man turret armed with a 7·62mm and 14·5mm machine gun.

When the West German Bundeswehr was formed it was forced to rely exclusively on imported equipment for several years before its own designs were ready to take the field. When they did, several of them were comparable in appearance and concept to AFVs with which the German Army had ended World War II although, of course, there was a marked improvement in performance. This was especially true in the case of the amphibious Spähpanzer Luchs (Lynx), which appeared in 1975 and has a similar suspension to the old SdKfz 234 armoured car series, which was in turn descended from the original ARW-MTW. The Luchs weighs 19,500kg (42,990lb), is manned by a crew of four, and is driven by two large propellers when afloat. It is armed with one 20mm cannon and a 7·62mm machine gun in a ground/air mounting, the turret being designed by Rheinmetall. Despite its size and weight, the vehicle is remarkably quiet on the move, a most important consideration during reconnaissance.

The postwar French Army found itself in a similar situation and for a while relied on British and Amer-

Top left: The Commando V-150. Centre left: The Commando Scout. Left: The Commando V-300. The Commando series of vehicles has proved to be very popular with countries that want an inexpensive and flexible AFV on a limited budget. It is in service with, among others, the armies of Bolivia, Malaysia, Turkey and Saudi Arabia.

ican wheeled AFVs and such pre-war designs of its own as had survived in France's colonial territories, the latter including the rear-engined four-wheel drive Panhard Model 178, which dated from 1935 but which was justly regarded as being advanced in its day. Another remarkable design of the period was the eight-wheel drive Panhard Model 201, on which work commenced in 1937. By 1940 a prototype existed: this was shipped to French North Africa to prevent its capture, and the plans were destroyed. The car itself was lost in the Sahara but in 1945 the ideas contained in its design were revived in the Panhard Model 212, which is better known as the EBR (Engin Blindé de Reconnaissance). Production of this symmetrical design began in 1950. The vehicle can be driven from identical positions at either end, and is powered by a 12-cylinder horizontally opposed Panhard 12 H 6000 engine which is only 220mm (8·5in) high and is housed under the floor of the fighting compartment. The front and rear wheels are fitted with conventional pneumatic rubber tyres, but the centre four are large metal discs, heavily flanged and with deep grousers, and can be raised when running on surfaced roads. Early marks were armed with a 75mm gun, most commonly mounted in the same oscillating FL-10 turret as the AMX13 light tank, but the vehicle was up-gunned in the late 1960s with a 90mm gun firing fin-stabilised ammunition. The car is manned by a crew of four and has a maximum speed of 105km/h (65mph). It saw much active service during the war in Algeria, as did an APC version, the EBT-ETT. The French are now producing a replacement for the Panhard EBR known as the AMX-10RC which uses many of the same components as the AMX-10P mechanised infantry combat vehicle. It weighs 15,000kg (33,075lb), has a top speed of 85km/h (53mph) and is armed with a 105mm MECA gun.

The French experience in Algeria led to the de-

Above: The Spähpanzer Luchs demonstrates its wintertime cross-country performance. The crew comprises two drivers (one at the front, the other at the rear), a commander and a gunner.

Above left: The Panhard AML with the H-90 turret. It also comes in three variants with a 60mm mortar, one with a 30mm cannon, plus AA and anti-tank versions. Above right: This Panhard EBR is in service with the Moroccan Army; the EBR has steering on the front and rear wheels.

Left: A BRDM-2 on an exercise. Right: A BRDM-1 with a PT76 in the background. The Soviet Army does not expect its reconnaissance units to fight and therefore provides them with small, lightly-armed and fast vehicles that try to avoid combat.

Below: Eland Mk 5s of the South African Army in Namibia. Bottom: A Brazilian Engesa Cascavel EE-9 armoured car fitted with an ENGESA ET-90 turret and EC-90 90mm gun.

velopment of the Auto-Mitrailleuse Légère, or AML Panhard Model 245. This light armoured car is similar to the British Ferret, with which it has competed very successfully in the arms markets of the world since 1961. With such sales potential in mind, Panhard have developed a variety of turrets and armaments to suit their customers' needs, including versions mounting a 90mm gun, a 60mm mortar, 30mm automatic cannon and several ATGWs; a licence to produce the AML has been granted to the South African company Sandock Austral Ltd which uses the name Eland for the vehicle. The basic AML weighs 5500kg (12,125lb), has a crew of three and a maximum speed of 100km/h (62mph); it is air-portable and can be lifted by the larger helicopters. The Argentinian Army shipped a squadron of 90mm AML 90s to the Falkland Islands during the 1982 conflict but these did not operate outside Port Stanley and took little or no part in the fighting.

More recently a few other countries have also produced their own armoured cars, but these tend to be variations on existing themes or designed solely for internal security duties. In general, since 1945 it has been the former colonial powers which have made the greatest use of the armoured car, in the counter-insurgency role where the use of tanks would have been uneconomic, if not undesirable. In this area, the armoured car has a marked advantage over tracked AFVs of every kind since it denies the insurgents the propaganda coup of claiming that the security forces are using 'tanks' and therefore excessive force against the population – an important consideration, of which the Soviet Union has failed to take adequate account in its handling of disturbances among its allies. In this light, it is easy to understand why the armoured car has proved to be such a valued tool in the hands of successive UN peacekeeping missions. Again, being cheaper to buy and maintain than tanks, armoured cars are the ideal AFVs for Third World nations with strictly limited defence budgets. As to their use in any major conflict, they now possess greater mobility and firepower than ever before, but the extent of their involvement will depend upon the nature of the fighting.

WESTERN MICVs

The West has lagged some way behind the Soviet Union in the design, development and deployment of mechanised infantry combat vehicles. The Soviets first fielded their BMP-1 in the mid-1960s but it was not until 1970 that West Germany introduced its Marder, and the Americans were even later with their M2 Bradley in 1983. The MICV has greater armour protection, mobility and firepower than the older armoured personnel carrier (APC), and the infantry have some provision to aim and fire their rifles from within the protection of the vehicle. They also normally have a full range of passive night vision equipment and a complete NBC system.

There is an important though sometimes blurred distinction between the APC (armoured personnel carrier) and the MICV (mechanised infantry combat vehicle) or IFV (infantry combat vehicle). The main role of the APC is to transport infantry from one part of the battlefield to another where they then dismount and fight as ordinary infantry. The MICV, however, is designed to operate with the forward armoured units as part of a combined arms team that includes tanks, self-propelled artillery, engineers and helicopters. The tactical requirements of the MICV are demanding: it has to have reasonable armour protection, have a good power-to-weight ratio in order to keep up with the tanks across rough terrain, be provided with a turret-mounted armament, have some provision for the infantry to use their smallarms from within the vehicle, have an NBC (nuclear, biological, chemical) warfare system and a full suite of passive night vision equipment for the driver, commander and gunner. Some countries also insist that the MICV must have an anti-tank capability as well, for example, the US Bradley IFV, and the West German Marder.

The first country to operationally deploy an MICV was not, however, in the West but in the East. The Soviet Union first displayed the BMP-1 MICV in 1967 and the introduction of this vehicle into service within the Soviet Army acted as an important spur to the West to accelerate the development of a similar vehicle.

When the West German Army was reformed during the 1950s requirements for new vehicles were formulated. The need for an APC was so urgent that a modification of an existing vehicle was chosen and this was rushed into service in 1958 as the Schützenpanzer 12-3. By 1960, however, the West German Army had a clearer idea of what it wanted and in that year it awarded contracts for prototypes of a MICV to be built, followed by pre-production vehicles, which were delivered in 1967-68. Following troop trials this vehicle was accepted for service as the Marder with the first production batch being completed in 1970.

The Marder has an excellent power-to-weight ratio, and armour protection over the frontal arc is sufficient to withstand penetration from 20mm projectiles. The two-man turret is fitted with an externally-mounted 20mm cannon with a co-axial 7.62mm machine gun. Many Marders have a Milan (*missile d'infanterie léger anti-char*) ATGW (anti-tank guided weapon) launcher mounted on the right side of the turret which gives it a useful anti-tank capability. Mounted over the troop compartment at the rear is a remote-controlled 7.62mm machine gun, while within the hull sides two firing ports a side have been built. The six infantrymen enter and leave the vehicle via a power-operated ramp in the hull rear.

Previous page: The US Bradley MICV thunders across desert terrain at speed. Above: The West German Marder is an expensive though highly effective MICV and it allows troops to use hull-mounted firing ports or to fight from the deck hatches (top). Below: Besides a 20mm cannon and two machine guns this Marder mounts a Milan anti-tank guided missile on the turret.

The French AMX VCI IFV, which entered service with the French Army in 1957-58, was one of the first postwar vehicles to have firing ports for the infantry and the option of a fully enclosed weapon station. It is now being replaced in the French Army by the AMX-10P which is fully amphibious and fitted with a two-man power-operated turret armed with a 20mm cannon and a co-axial 7.62mm machine gun. The rear troop compartment does not have any provision to aim and fire weapons from within the vehicle apart from two firing ports in the power-operated ramp in the hull rear. There are many variants of the AMX-10P including an anti-tank vehicle with four HOT (high-subsonic optically-tracked tube-launched) ATGWs, command vehicle, mortar towing vehicle, ambulance and various artillery and observation vehicles. For the export market the AMX-10 PAC 90 has been developed which is the basic AMX-10P fitted with a two-man GIAT TS-90 turret armed with a long-barrelled 90mm gun. This vehicle, already in service with the Indonesian Marines, is designed to provide fire support for other AMX-10P vehicles.

The United States Army saw the need for an MICV in the early 1960s and the first vehicle to meet this requirement was the XM701 which used parts of the M110 and M107 self-propelled guns and was fitted with a turret-mounted 20mm cannon, but this did not progress beyond the prototype stage. Based on the M113 APC the FMC Corporation built the XM765 which had firing ports in the rear troop compartment and a new weapons station. This was not selected for service, however, although further development some years later by FMC as a private venture resulted in the AIFV (armoured infantry fighting vehicle) which was selected by the Netherlands in 1975. The AIFV was later ordered by the Philippines and in 1979 Belgium ordered 514 vehicles to be built in the country under licence. For many countries who cannot afford the highly complex Bradley IFV, the AIFV is an ideal compromise. Over the earlier M113 the AIFV offers improved armour protection, greater firepower (25mm cannon against a 12.7mm machine gun) and better cross-country mobility owing to improved suspension.

Above: French Army troop carriers – the AMX VCI (top), the AMX 10P with troops, weapons and equipment (above left) and an AMX 10P swimming a water obstacle armed with a 20mm cannon (above). Below: The US AIFV (with TOW launcher) in Dutch Army service. Below left: Troops fire from an AIFV.

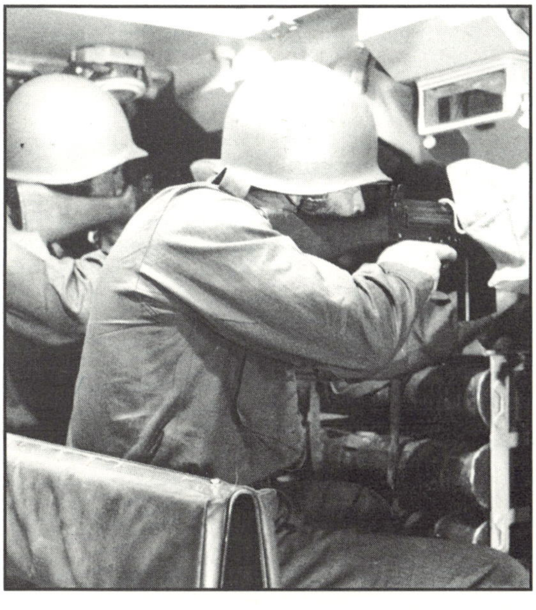

Late in 1972 FMC were awarded a contract to design and build prototypes of a new MICV under the designation of the XM723, with all 12 prototypes to be completed by 1975. At that time prototypes of a new ARSV (armoured reconnaissance scout vehicle) had been built but a decision was taken to cancel this project and build a vehicle to meet both standard MICV and scout requirements. At the same time it was decided that both vehicles would have a new two-man turret (commander and gunner), be armed with a 25mm cannon and have a twin launcher for the Hughes TOW (tube-launched optically-tracked wire-guided) ATGW to give the vehicle a long range anti-tank capability. The infantry fighting vehicle was designated the M2 while the scout or cavalry fighting vehicle became the M3. In 1977 it was decided to use the same chassis for what is now known as the MLRS (multiple launch rocket system). In 1980 the United States Army ordered 100 M2/M3 vehicles and all of these were completed by July 1982, and subsequently were produced at the rate of 50 vehicles per month. The M2/M3 was named the Bradley and became operational in 1983 in the United States and the following year in West Germany. The United States Army has a requirement for nearly 7000 Bradley MICVs but these will not replace the M113 on a one-for-one basis as in many battlefield roles – such as command post, mortar carrier, logistic support and ambulance – the older M113 is more than adequate.

The M2 has a three-man crew – commander, gunner and driver – and it can carry up to seven infantrymen in the rear compartment. The hull is constructed of welded aluminium with spaced laminate armour added to the hull front, sides and rear for extra protection. All the vehicles are fully amphibious and rely on their tracks for propulsion.

The 25mm Hughes dual-feed Chain Gun is mounted in a power-operated turret which has a stabilisation system that allows the cannon to be aimed and fired whilst the vehicle is moving across country at speed. A total of 300 rounds of ammunition are carried for

Top left: A side view of an M2 Bradley revealing its turret-mounted 7.62mm machine gun and 25mm Hughes Chain Gun. Above: Advancing up a river bed an M2 demonstrates its water-crossing ability. The M2 also has a full amphibious capability; it is able to float and propulsion is provided by the action of its tracks.

Above: The 'cavalry' version of the Bradley, the M3, whose main function is to act as a battlefield reconnaissance vehicle. As with the infantry M2, the M3 has a twin TOW launcher mounted on the left side of the turret. Above right and right: The British MCV-80, which was designed to supplement and extend the capabilities of the FV432 APC. Despite a number of developmental problems the MCV-80's 30mm main armament and good armour protection make it a potent battlefield weapon.

immediate use with a further 600 rounds kept in reserve. Three types of 25mm ammunition are currently available: APDS-T (armour-piercing discarding-sabot – tracer), HEI-T (high-explosive incendiary – tracer) and a training round. Under development is an APFSDS (armour-piercing fin-stabilised discarding-sabot) round with greatly improved penetration. Mounted co-axially with the 25mm cannon is a 7.62mm machine gun, while mounted on the left side of the turret is a twin launcher for the Hughes TOW ATGW, with a further five TOW missiles stowed in the hull rear. A bank of four electrically-operated smoke dischargers is mounted either side of the turret and latest production vehicles have an engine smoke generation system along similar lines to that fitted in Soviet tanks. The M3 cavalry fighting vehicle is almost identical to the M2 but has no firing ports; it has a three-man crew and carries only two infantrymen who can dismount from the vehicle to carry out reconnaissance on foot. The M3 has a greatly increased ammunition carrying capacity compared with the M2 as it is envisaged that on occasions it will have to fight to obtain information.

The United Kingdom's MICV is called the MCV-80 with the first 250 being ordered in 1984. The MCV-80 has a three-man crew – commander, gunner and driver – and can carry seven fully-equipped infantrymen. An unusual feature of the MCV-80 when compared to other MICVs is that there is no provision for the infantrymen to fight from within the vehicle as this is not a British Army requirement. The MCV-80 is fitted with a two-man power-operated turret armed with a Rarden 30mm cannon (as fitted to the Scimitar and Fox reconnaissance vehicles) and a co-axial 7.62mm machine gun. It has no anti-tank capability, however, as the British believe that this role is best left to dedicated vehicles such as the Striker. The MCV-80 has a full range of passive night vision equipment and an NBC system, but no amphibious capability. To give it the latter the vehicle would have to be lighter and armour protection would therefore have been lessened.

The standard APC of the Italian Army is the American M113A1 which has been manufactured under licence in Italy by OTO-Melara. Further development of the M113A1 resulted in the AIFV (armoured infantry fighting vehicle) which has applique armour added to the hull front, sides and rear, new fuel tanks on either side of the power-operated ramp at the hull rear (thereby increasing space in the troop compartment as well as reducing the fire risk), sloping upper sides to the troop compartment with vision blocks and firing ports underneath, and armament comprising 12.7mm and 7.62mm machine guns.

OTO-Melara is currently developing the VCC-80 infantry fighting vehicle to meet the requirements of the Italian Army. This will have a low profile and be fitted with a power-operated turret armed with a 25mm cannon and a 7.62mm co-axial machine gun. At one time it was proposed to fit Milan ATGWs to the side of the turret but it is now expected that a dedicated anti-tank version will be developed instead.

For the export market the company have recently developed the OTO C13 which has firing ports in the rear troop compartment and a variety of weapon stations ranging from a simple 12.7mm machine gun up to a turret-mounted 20mm cannon or even a 90mm gun. Its main advantage over the M113 is its much lower profile and improved armour protection.

MOWAG of Switzerland built some of the first prototypes of the Marder MICV for the West German Army in the early 1960s and since then MOWAG has undertaken a continuous programme of private venture research and development which has culminated in the MOWAG Improved Tornado MICV. This can be fitted with a wide range of armament installations including a new Oerlikon-Bührle two-man power-operated turret armed with a 25mm cannon, 7.62mm machine gun and twin launcher for TOW ATGWs. At present the Swiss Army operates M113A1 series APCs but has a requirement for a new MICV which could be met by the Improved Tornado MICV.

While the major powers are now developing or employing mechanised infantry combat vehicles there are wide differences between many of the vehicles in terms of armour protection, mobility and firepower. In some cases this is because the role of the MICV/IFV on the battlefield is seen differently by the various armies employing them but in other cases the vehicle has had to be developed within clearly defined cost constraints which has meant that many features have been left off.

As the 1973 Middle East war proved, tanks cannot survive on the modern battlefield without the support of other arms, including infantry in mechanised infantry combat vehicles. There are, however, a number of observers who believe that the MICV will not survive on the battlefield alongside the tank unless it has a similar degree of protection. This would mean MICVs weighing at least 40 tonnes but this would make the vehicles even more expensive which in turn would make them unacceptable to most armies.

Below: The Swiss MOWAG Tornado MICV, a well-armed fighting vehicle broadly similar to the West German Marder. Centre and bottom: Two views of the Italian OTO C13, armed with a turret-mounted 0.5in Browning machine gun. Built for the export market the OTO C13 is a rugged and flexible fighting vehicle offering superior performance over conventional APCs.

WESTERN APCs

The United States was the first Western country to field an armoured personnel carrier after the end of World War II, but early models such as the M75 and M59 were expensive to build, maintain and operate. They were followed by the M113 which has become the most widely used armoured fighting vehicle in history. Production commenced in 1960 at FMC's plant in California and by 1984 well over 80,000 had been built. Like most other APCs the basic vehicle has been adopted for a wide range of roles, many of which were not even thought of at the design stage. Britain, Austria, Belgium, Sweden and France have all produced APCs since 1945 but France is the only country to have succeeded in exporting them on a large scale.

The Americans and the Germans made the most extensive use of APCs (armoured personnel carriers) during World War II, but both countries used half-track vehicles which had limited cross-country mobility compared with the full-tracked tanks with which they were designed to operate. The half-tracks also suffered the major disadvantage of having an open-top troop compartment which made the infantrymen very vulnerable to artillery and shell splinters. Towards the end of World War II the British and Canadians removed the turrets from tanks and successfully used them in a troop carrying role, some remaining in service well after the end of the war. These tank conversions also suffered from having an open-topped troop compartment – although they could operate with the tanks – and the troops they carried had to climb over the hull sides to enter and leave the vehicle.

The United States Army was the first to field a new fully-tracked APC, followed by the French and Germans in the 1950s and the British in the 1960s. Since then a number of other countries have entered the field. In many cases these APCs have been sold overseas, so helping to extend production lines and preserve jobs in the defence industries.

The major difference between the APC and the MICV (mechanised infantry combat vehicle) or IFV (infantry fighting vehicle) is that the first is designed to transport the infantry and its equipment from one part of the battlefield to another where they dismount and fight on foot, while in the case of the MICV the infantry will usually fight from within the vehicle. When first introduced most APCs had a pintle-mounted machine gun but experience in South Vietnam and elsewhere has shown that the gunner was very vulnerable and many countries have replaced the pintle-mounted weapons with a similarly armed turret. Most APCs have no capability to allow the infantry to fight from within the vehicle with any degree of safety, unless they expose the upper part of

Page 101: Armed with 20mm M61A1 Vulcan cannons a line of US M113 armoured personnel carriers moves forward on exercises in Germany. Above left: An M113 rejoins the road after travelling cross-country. The ability of APCs to transport infantry over rough terrain while providing protection from smallarms is a crucial element in modern infantry tactical doctrine. Above: The M106A1 mortar carrier. Based on the M113 it carries 88 rounds of 107mm mortar ammunition, some of which can be observed racked behind the mortar.

Left centre: America's first fully tracked APC, the M75, seen here with a 0.5in machine gun. Left: The M59 was a considerably more advanced APC than the M75; troops could leave the vehicle with ease through a power-operated rear door.

their body through the roof hatches; and most APCs also have no observation devices in the troop compartment, so prior to dismounting the infantry have no clear idea of the terrain in which they are expected to fight.

Some APCs are fully amphibious needing little or no preparation, while others are only amphibious with some preparation or have no amphibious capability at all. This capability nevertheless only allows them to cross lakes or slow-flowing rivers; it does not mean that they can swim from landing craft offshore onto beaches. When most APCs were introduced they were not fitted with a NBC (nuclear, biological and chemical) warfare system to enable them to operate in a NBC contaminated area, but within Nato many vehicles have now fitted these systems. The original infra-red night vision equipment has now given way to passive night vision, which allows the vehicle to be driven in the dark without giving away its position to other infra-red sensors, as well as being able to dispense with standard white-light headlights which can be seen by an enemy for miles around.

In most cases the basic APC has also been used to form the basis of a complete family of vehicles that can support it on the battlefield: for example, mortar carrier, command, artillery observation/fire control, engineer support, ambulance and radar carrier to name but a few. Some are even used as weapons platforms, fitted with anti-aircraft or anti-tank systems.

While the MICV is a far more comprehensive combat vehicle than the APC, it is also much more complex and expensive. For example, in 1983 an American M2 Bradley MICV cost the American Army $1.4 million, while the M113A2 APC cost only $176,000.

Of all the nations producing APCs the most important has been the United States, producing an extensive range of vehicles. Following their early experience with half-tracked APCs the American Army soon issued a requirement for a fully-tracked fully-enclosed vehicle and this eventually emerged as the M75 APC with just over 1700 being built between 1951 and 1954 by International Harvester. This was, however, a very large vehicle and had no amphibious capability so it was soon replaced in the US Army by the M59. Almost all of the M75s were then passed on to Belgium where they remain in service.

The M59 was designed by the FMC Corporation who first became involved in the design and production of armoured vehicles during World War II. Between 1954 and 1959 over 4000 M59s were built by the company for the US Army. The M59 was a major improvement over the M75 as not only was it fully amphibious but infantry could quickly leave the rear troop compartment through a power-operated ramp in the hull rear. Once the M59 was phased out of US Army service in the early 1960s it was passed on to other countries such as Brazil and Greece where it still remains in service.

The main drawback of the M59 was that it was a little on the heavy side for air transport and it had a short operating range. To meet a requirement for a lighter vehicle, prototypes of a new APC were built both in steel and aluminium and the latter was finally adopted for service as the M113 with prime contractor being FMC once again. First production vehicles were completed in 1960 and since then it has become the most widely used armoured vehicle in the world. By 1984 sales had been made to some 50 countries and over 80,000 vehicles had been built; licensed production has been undertaken in Italy and Belgium. The first model was powered by a petrol engine but in 1964 the M113 was replaced by the diesel-engined M113A1 which has a much improved operating range. The current production model is the M113A2 which has further automotive improvements and the fuel tanks mounted externally at the rear which not only saves space inside but reduces the fire risk.

There are more variants of the M113 family than any other armoured vehicle; these include 81mm and 107mm mortar carriers, an anti-tank vehicle with the TOW ATGW (anti-tank guided weapon), command vehicle (with higher roof), flamethrower, 20mm anti-aircraft vehicle, recovery and repair vehicles and a cargo carrier called the M548 which in turn is used for a variety of roles including Lance and Chaparral missile carriers.

The main drawback of the M113 is that there is no provision for the infantry to use their weapons from within the vehicle with any degree of safety. This has led FMC to develop the AIFV (armoured infantry fighting vehicle) which is mid-way between the M113 and the M2 Bradley. The AIFV has combined aluminium and steel armour and a turret-mounted 25mm cannon. In service with the Netherlands and Philippines, it has been ordered by the Belgian Army.

After the end of World War II the United Kingdom started a major development effort to produce a fully-tracked APC, although it was the six-wheeled Saracen that entered production first as this was urgently required for operations in Malaya. The FV432 series was produced by GKN Sankey with over 3000 vehicles being produced between 1963 and 1971. In many respects this is very similar to the American M113 except that it has a welded steel hull rather than a welded aluminium hull. There is no provision for the infantry to use their weapons from inside the vehicle with any degree of safety and when first introduced it was armed with a pintle-mounted 7.62mm machine gun. Many vehicles now have a fully enclosed 7.62mm machine-gun turret over the rear troop compartment which is a far more satisfactory arrangement. Some of the vehicles used by the Berlin Brigade have the complete two-man turret of the Fox armoured car armed with a 30mm Rarden cannon which is capable of destroying any Soviet MICV/APC as well as inflicting severe damage to the sides and rear of a main battle tank.

The chassis of the FV432 has been used for a wide range of specialised variants to meet not only the requirements of the infantry but other arms as well. These variants include a command vehicle with a tent that can be erected at the rear of the hull, 81mm mortar carrier, recovery vehicle, ambulance and cargo carrier. More specialised versions include the FV434 maintenance carrier that has a hydraulic crane for changing tank engines in the field, radar carrier (including the EMI Cymbeline mortar-locating radar), FV438 anti-tank vehicle with two Swingfire ATGWs and a variety of Royal Signals and Royal Engineers versions. Included in the latter is a variant that tows the bar minelayer while mounted on the roof is the Ranger anti-personnel mine dispensing system. The FV433 105mm Abbot self-propelling gun, used by the Royal Artillery and India is also based on the chassis of the FV432 APC.

The FV432 was offered on the world market but no sales were made as the American M113, which has similar characteristics, is so much cheaper. From the late 1980s the FV432 will be supplemented by the MCV-80, which is also being designed and built by GKN Sankey.

There is an APC version of the Alvis Scorpion family called the Spartan which is used by the British Army, Royal Air Force Regiment and the Belgian Army. This vehicle has a three-man crew – comman-

Britain has produced three fully-tracked APCs which provide a comprehensive series of infantry support vehicles. The FV432 (below) is the standard British Army APC but will be supplemented by the MCV-80 in the late 1980s. The Alvis Stormer (bottom left and right) has been built specifically for the export market and is already in service with the Malaysian Army. The Stormer is a larger-hulled member of the Scorpion family of vehicles.

der, gunner and driver – but carries only four infantrymen and so is a very limited APC. To meet the requirement for an APC with increased carrying capacity Alvis have developed the Stormer which was announced in 1980. The Stormer can can carry eight fully-equipped infantrymen in addition to its three-man crew and can be fitted with a wide range of weapon stations. Alvis supplied three Stormers for the American Light Armored Vehicle competition, fitted with a two-man power-operated turret armed with a 25mm Hughes Chain Gun and firing ports in the rear troop compartment, and more recently has supplied 25 vehicles to Malaysia, which also operates the Scorpion.

France was one of the first European countries to issue a requirement for a full tracked APC in the postwar period and after evaluating prototypes from several companies finally selected a model based on the AMX13 light tank chassis. This was originally called the TT CH for short, but is today known as the VCI (*véhicule de combat d'infanterie*). This was one of the first vehicles to have firing ports for the infantry, although it did lack any amphibious capability. When originally introduced it was armed with a pintle-mounted 7.62mm machine gun but most have now had this replaced by a turret armed with a similar weapon, although 20mm weapon stations can be installed if required. In addition to being used by France it has also been exported to Argentina, Belgium, Ecuador, Indonesia, Italy, Morocco, the Netherlands, Venezuela and the United Arab Emirates. In the French Army it is slowly being replaced by the AMX-10P MICV which, in addition to having a power-operated 20mm turret, is also fully amphibious.

For many years the Belgian Army has been equipped with the French AMX VCI and old American M75 APCs. After evaluating a number of tracked and wheeled vehicles the army selected both the American M113A2 and the AIFV in 1979. Rather than purchase the vehicles direct it was decided to make them under licence in Belgium. First production vehicles were completed in 1982 and production will continue to 1988. The Belgian company of ACEC has developed the Cobra armoured personnel carrier, which has an interesting range of armament comprising two 101mm rocket launchers, a 0.5in machine gun, two 7.62mm machine guns and two triple grenade launchers. An unusual feature of this vehicle is that it has an electric transmission.

To give Swedish infantry some experience of armoured personnel carriers Hägglund and Söner converted a number of light tanks into APCs and called them the Pbv 301. These were later replaced by the Pbv 302 designed by the same company. This vehicle entered production in 1965 and in many respects was ahead of the American M113 and British FV432 (introduced at about the same time) as it was fitted with a fully enclosed turret armed with a 20mm cannon. The driver is positioned at the front of the vehicle in the centre and the commander to the right. The latter has periscopes to give good all-round observation. Over the troop compartment roof are two hydraulically-operated roof hatches which allow the troops to fire their smallarms but with their heads and shoulders exposed. As there are so many lakes and rivers in Sweden the Pbv 302 is fully amphibious being propelled in the water by its tracks at a speed of 8km/h (5mph). The hull of the Pbv 302 is of double-

Top: The Italian modification of the M113 which provides firing ports for the infantry in the rear hull, becoming in effect an infantry combat vehicle. Above centre: The French APC, the VCI, based on the AMX13 chassis. Although it is being replaced by the AMX-10P MICV it has been an important export success. Above: The Swedish Pbv 302 with 20mm cannon and infantry firing position. Right: The improved Pbv 302 Mk2 which has stronger armour and an extra cupola.

Western Armoured Personnel Carriers

Vehicle	Crew	Weight	Ground Pressure	Speed (road)	Range (road)	Armament
Steyr 4K 7FA	2+8	14,800kg (32,634lb)	0.55kg/cm² (7.82lb/in²)	63.6km/h (39.2mph)	520km (323 miles)	one 12.7mm machine gun
Cobra	3+9	7500kg (16,537lb)	0.29kg/cm² (4.12lb/in²)	80km/h (50mph)	600km (373 miles)	one 12.7mm machine gun
AMX VCI	3+10	15,000kg (33,075lb)	0.70kg/cm² (9.95lb/in²)	65km/h (40.4mph)	400km (248 miles)	one 7.62mm/7.5mm machine gun
Pbv 302	2+10	13,500kg (29,767lb)	0.60kg/cm² (8.53lb/in²)	66km/h (41mph)	300km (186 miles)	one 20mm cannon two × 4 smoke grenade launchers
FV432	2+10	15,280kg (33,692lb)	0.78kg/cm² (11.09lb/in²)	52km/h (32mph)	480km (298 miles)	one 7.62mm machine gun two × 3 smoke dischargers
Stormer	3+8	10,700kg (23,593lb)	0.37kg/cm² (5.26lb/in²)	72km/h (45mph)	800km (497 miles)	one 7.62mm machine gun or 20/30mm cannon, 76/90mm gun, ATGW depending on role
M113A1	2+11	11,156kg (24,600lb)	0.55kg/cm² (7.82lb/in²)	68km/h (42mph)	480km (298 miles)	one 12.7mm machine gun

skinned construction which also provides protection against HEAT attack as well as additional buoyancy. The Pbv 302 has not been sold abroad owing to Sweden's strict policy against the export of arms. More recently the Pbv 302 Mk 2 has been developed with improved armour protection, Lyran target illumination launchers and an additional cupola.

The standard APC of the Swiss Army is the M113, although many of these have been modified to meet Switzerland's own requirements. To provide increased firepower some have also been fitted with the same 20mm cannon and turret as installed on the Swedish Pbv 302.

The first APC designed to meet the requirements of the reformed West German Army was the HS-30. This was based on a Swiss-designed chassis, powered by a Rolls-Royce petrol engine and armed with a turret-mounted 20mm cannon. It has now almost disappeared from service, having been replaced by the M113 and the Marder MICV.

In the 1960s Saurer-Werke built almost 500 fully-tracked armoured personnel carriers for the Austrian Army under the designation of the 4K series, some of which were fitted with pintle-mounted machine guns while others had a turret-mounted 20mm cannon. In 1970 the company was taken over by Steyr-Daimler-Puch and subsequently a more powerful version was built under the designation of the 4K 7FA and sales of this have already been made to Austria, Greece (where it is now made under licence), Morocco, Nigeria and Tunisia. More recently a model called the 4K 7FA-K Spz infantry fighting vehicle has been built with firing ports in the rear troop compartment. Variants include a mortar carrier, ambulance, command, anti-aircraft and a fire support vehicle with a 90mm gun.

Rather than design its own APC Italy made the sensible decision to build the M113 under licence. Prime contractor for this is OTO-Melara at La Spezia who also build tanks and a wide range of naval weapons. So far over 4000 vehicles have been built for the Italian Army and export and further development has resulted in an infantry fighting vehicle which has firing ports in the rear troop compartment and a partly-closed weapon station.

Although mechanised infantry combat vehicles are being introduced in increasing numbers in many armies the basic armoured personnel carrier still fulfils its original mission of transporting troops into battle. Far more economical than the highly complex MICV, the armoured personnel carrier is able to carry out a wide range of support roles, and it will remain a battlefield vehicle for many years to come.

Below: The Steyr 4K 7FA which is armed with a turret-mounted 20mm cannon and is capable of holding eight infantrymen in the hull. Bottom: The Belgian Cobra, shown here without its rocket launchers. Its light weight of 7500kg (16,537lb) is an advantage in many respects but ensures only limited protection.

SOVIET APCs

The Soviet Army deploys two types of armoured personnel carrier, wheeled and tracked. The former offers greater strategic mobility as it is faster on roads but it is only lightly armed and armoured. The tracked BMP-1 and BMP-2, however, have better cross-country mobility, better armour protection and enough firepower to engage every type of target on the battlefield, including enemy tanks. The unique BMD airborne combat vehicle can be dropped by parachute behind enemy lines, where it can create havoc as it has the same turret as the BMP-1. The more recent MT-LB is a multi-purpose vehicle, undertaking various roles such as carrying troops and stores, hauling anti-tank guns, and serving as a command vehicle.

Since the Second World War the Soviets have deployed large numbers of Armoured Personnel Carriers (APCs) to give their infantry a measure of protection and the speed to keep up with a typical armoured assault. In November 1961 a new vehicle appeared which represented a considerable advance on previous Soviet APCs. The BTR-60, a 10-tonne vehicle, has eight powered wheels with power-assisted steering on the front four and two 90hp petrol engines that provide a road speed of 80km/h (50mph). It is fully amphibious with a single hydrojet for propulsion and two smaller jets for steering and is capable of a water speed of 10km/h (6mph). The boat-shaped welded steel hull has a good ballistic shape and 'swimming' characteristics, while torsion bar suspension with hydraulic shock absorbers endow it with a high degree of cross-country mobility.

The original BTR-60P model was open-topped, but overhead protection and an NBC system were introduced in the BTR-60PA variant. The most common version, the BTR-60PB, incorporated a small machine gun turret housing a 14.5mm KPV heavy machine gun and a secondary 7.62mm PKT co-axial machine gun. The driver is seated front left with the vehicle commander on his right, and both are provided with direct vision windscreens protected by steel hatches that can be raised and infra-red driving lights and searchlight for night operations.

The BTR-60PB can carry up to 14 infantry in addition to the two crew members. One of the main disadvantages of its design, however, is that it has no rear exit doors. Mounted troops have to climb out of

Previous page: A battle-scarred BTR-60PB lies abandoned after action in Golan during the 1973 Middle East war. Right: Infantry dismount from the early model BTR-60P while one of the vehicle's SGMB 7.62mm machine guns provides covering fire. This open-topped model was soon replaced by the PB variant which provided overhead armour, but dismounting through roof hatches still remained hazardous.

Below: A column of Soviet naval infantry BTR-60s rolls ashore from a landing ship. The lead vehicle is a BTR-60PB which mounts the distinctive small conical turret housing the 14.5mm HMG main armament. Clearly visible on the hull side are the numerous hand and footholds which allow the infantry complement to mount and dismount from the vehicle.

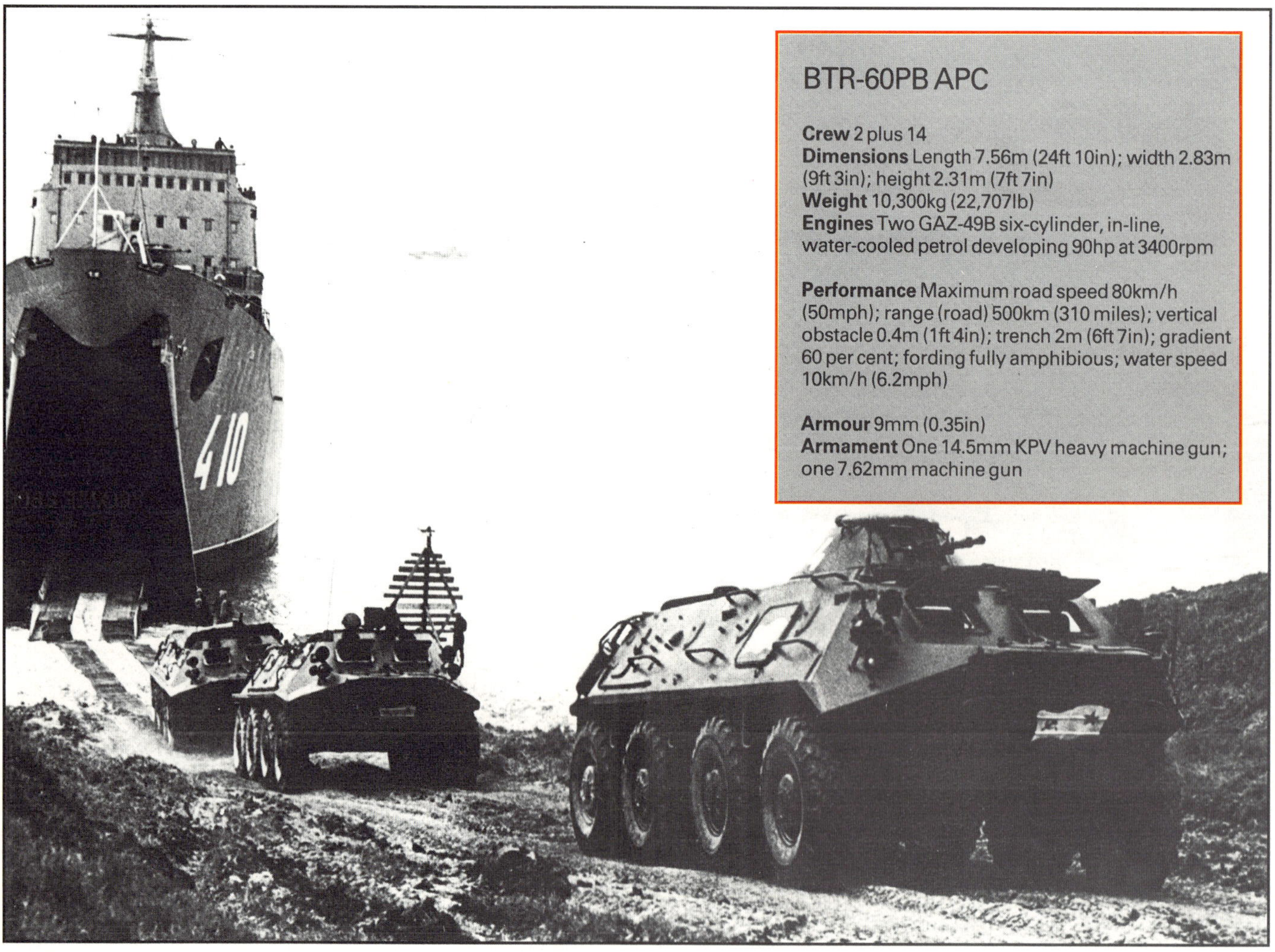

BTR-60PB APC

Crew 2 plus 14
Dimensions Length 7.56m (24ft 10in); width 2.83m (9ft 3in); height 2.31m (7ft 7in)
Weight 10,300kg (22,707lb)
Engines Two GAZ-49B six-cylinder, in-line, water-cooled petrol developing 90hp at 3400rpm

Performance Maximum road speed 80km/h (50mph); range (road) 500km (310 miles); vertical obstacle 0.4m (1ft 4in); trench 2m (6ft 7in); gradient 60 per cent; fording fully amphibious; water speed 10km/h (6.2mph)

Armour 9mm (0.35in)
Armament One 14.5mm KPV heavy machine gun; one 7.62mm machine gun

Right: A BTR-70, the modified version of the 60PB, on display. While retaining the basic layout and armament of the PB, the 70 is provided with larger troop compartment hatches and improved service access to the two diesel engines. Other modifications include a redesigned wheelbase with larger tyres and a flatter shaped hull.

Below: With trim vane extended and hydrojet vent casing fully open, a BTR-60PB demonstrates its amphibious capability. Below inset: Afghan guerrillas in characteristic victory posture pose for photographers on a captured Soviet BTR-60.

the two small roof hatches and jump over the side of the vehicle, a hazardous operation while under heavy fire in close support operations. The BTR-60PB also has a volatile fuel system making it liable to 'brew up' if hit.

Despite its age and design inadequacies the BTR-60 still equips the majority of front line Soviet Army troops, as well as the Soviet naval infantry, and is a reliable and versatile vehicle. Further variants include the BTR-60PU command model and the BTR-60PK long-range communications vehicle. The BTR-60 equips a number of countries outside the Warsaw Pact and it was the main APC of the Arab armies in the Yom Kippur War of October 1973. It has also seen action with the North Vietnamese Army in Vietnam, with the Ethiopians in the Ogaden War (1977-78), and in Angola. In 1978 an improved version known as the BTR-70 made its appearance. It is fitted with more powerful petrol engines and has an improved troop compartment hatch arrangement.

When the BMP mechanised infantry combat vehicle (MICV) first appeared in 1967 it aroused considerable interest among Western analysts. It mounts a 73mm gun with a co-axial 7.62mm PKT machine gun and is provided with an AT-3 Sagger anti-tank guided weapon (ATGW) launcher with one missile mounted on the launch rail and four further rounds carried inside the vehicle. The combination of these weapons provides considerable firepower for the eight-man infantry section. The 73mm 2A20 smooth-bore main gun is not stabilised and is fed by a semi-automatic loading system. A major disadvantage is the tendency for this system to feed the loader's arm, instead of the shell, into the breech. Its slow rate of fire, eight rounds per minute, is largely attributable to the design which automatically elevates the gun during the reloading cycle.

The BMP carries 40 HEAT and HE rounds. The PG-9 fin-stabilised projectile, which resembles that of the RPG-7, is vulnerable to deflection by cross-winds. If the gunner is to have any real prospect of hitting his target the vehicle must be halted. Even so

BMP-1 MICV

Crew 3 plus 8
Dimensions Length 6.74m (22ft 1in); width 2.94m (9ft 8in); height 2.15m (7ft 1in)
Weight Combat loaded 13,500kg (29,762lb)
Engine One V-6 six-cylinder diesel developing 290hp at 2000rpm

Performance Maximum road speed 80km/h (50mph); range (road) 500km (310 miles); vertical obstacle 0.8m (2ft 7in); trench 2m (6ft 7in); gradient 60 per cent; fording fully amphibious; water speed 8km/h (5mph)

Armour 23mm (0.9in)
Armament One 73mm gun; one 7.62mm machine gun co-axial with main armament; one launcher rail for AT-3 Sagger ATGW

Right: Since the early 1950s Soviet APC design has placed increased emphasis on weight of armament. While early vehicles such as the BTR-152 carried light support machine guns, later generations were upgunned to provide heavier covering fire and attack capability. Armament on the BMP MICV shown here includes both a 73mm main gun with a co-axially mounted 7.62mm machine gun and the AT-3 Sagger ATGW system.

he is at some disadvantage since there is a blind spot caused by the infamous raised projection on the turret ring which automatically elevates the gun barrel to prevent it striking the commander's infra-red searchlight. For targets over 800m (875 yards), the Sagger missile is selected, but much dexterity is required to load, launch and hold the wire-guided missile on course. The maximum range of the Sagger system is 3000m (1.8 miles).

The vehicle's driver is seated front left with the commander behind him, while the infantry complement occupy the troop compartment to the rear. Infantry enter and exit through two doors in the rear of the hull and are also provided with four hatches and periscopes in the roof. Smallarms firing ports are located in the hull sides.

One of the main considerations in the BMP's design was mobility, allowing the vehicle to travel fast in an advance across an NBC-contaminated battlefield. This 14-tonne tracked APC is powered by a six-cylinder diesel engine and has a road speed of 80km/h (50mph). Like the BTR-60 it is fully amphibious, using its tracks for propulsion, and is fitted with an NBC defence system. Vulnerability is still a problem, however, due to its relatively thin armour. Experience in the Middle East wars has highlighted the risk of incineration of the crew should the fuel cells in the rear doors and passenger compartment or ammunition stowage be hit. Yet, used aggressively and in strength, the low-silhouetted vehicle poses a serious threat on the battlefield.

The BMP series also includes the BMP-2, which appeared in late 1982. This variant carries a 30mm machine cannon and the AT-5 Spandrel anti-tank missile system. Other variants are the BMP-R reconnaissance vehicle and the BMP-U command vehicle with additional communications equipment; the BMP-POO mobile training centre; BMP Radar with

Below: A formation of Soviet BMP-1s on exercise. While most Soviet-built APCs have a crew of two, the BMP's heavier armament necessitates the inclusion of a gunner in addition to the vehicle's driver and commander. The driver's position is situated at the front of the hull with the commander behind him while the gunner occupies the one-man turret. Both driver and gunner wear overalls and are armed with pistols while the commander wears battledress and carries an assault rifle.

Left: Smallarms firing ports and individual infantry periscopes on the hull of the BMP.

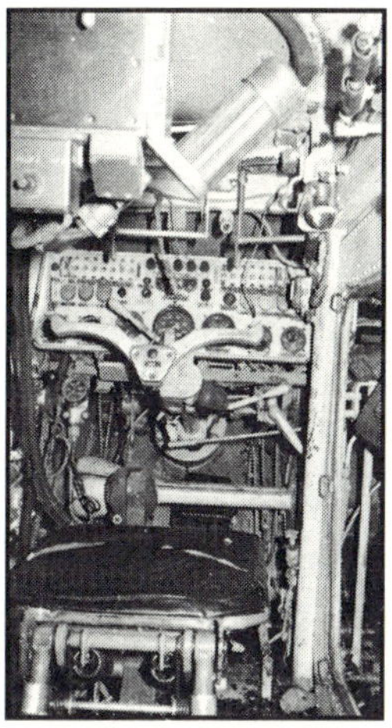

Right: Two views showing details of the interior arrangement of the BMP; the driver's position with T steering bar (left) and a section of the troop compartment (right). Mounted infantry are seated on benches facing outwards and are provided with twin rear doors for entry and exit from the vehicle.

Left: Rather than risk the possibility of incineration, an Iraqi crew beats a hasty retreat from their BMP during an Iranian air attack.

a two-man turret on the rear of which is a Small Fred radar; while Czechoslovakia has built an armoured recovery vehicle on the BMP chassis.

The BMD, which first appeared in 1973, is an APC designed for deployment with the Soviet airborne forces. Based on the BMP, but a much lighter vehicle, it can be air-dropped and carries a crew of three with six airborne infantry occupying open seats at the rear. The BMD is powered by a 280hp diesel engine and has a road speed of 80km/h (50mph). It is a fully amphibious vehicle, propelled by two hydrojets, and is provided with NBC fit and night-vision equipment. Armament consists of a 73mm gun, two 7.62mm machine guns in the bow, one co-axial with the main

Right: A captured BMP-1 on exhibition in Israel. Several Middle East Arab armies are equipped with this vehicle. In combat these BMPs have not always been deployed to their best tactical advantage and the 1973 war saw the Golan Heights littered with knocked-out Syrian APCs.

gun and the AT-3 Sagger ATGW system. Like so many Soviet APCs, the BMD has seen action in the Ogaden as well as being extensively deployed in Afghanistan.

A further addition to the Soviet APC inventory is the MT-LB multi-purpose tracked vehicle. It was originally designed as an artillery tractor but its role has been extended to include cargo transport, APC, command, communications and prime mover deployments. Like all later generation Soviet APCs it is fully amphibious and NBC protected, although it is only lightly armed with a 7.62mm machine gun.

Soviet tactical thinking places a great deal of emphasis on the deployment of APCs, which have a considerable role to play in both first and second echelon assaults. Crew and infantry protection, however, has not been the prime design consideration; and although the Soviet APCs achieve high cross-country mobility, they are very vulnerable to modern anti-armour battlefield weapons systems in the hands of highly-skilled operators.

Above left: Soviet motorised infantry disembark through the rear doors of a BMP. Above right: The smaller BMD which equips the Soviet airborne forces. In an assault without tank support, the BMD normally precedes the dismounted infantry and is provided with two bow-mounted machine guns for extra frontal firepower.

Below: Capable of being dropped by parachute, the BMD provides important battlefield and cross-country mobility for accompanying airborne troops.

BMD Airborne Infantry Combat Vehicle

Crew 3 plus 6
Dimensions Length 5.3m (17ft 5in); width 2.65m (8ft 8in); height 1.85m (6ft 1in)
Weight Combat loaded 9000kg (19,840lb)
Engine One V-6 water-cooled diesel developing 280hp

Performance Maximum road speed 80km/h (50mph); range (road) 320km (200 miles); vertical obstacle 0.8m (2ft 7in); trench 1.6m (5ft 3in); gradient 60 per cent; fording fully amphibious; water speed 10km/h (6.2mph)

Armour 6mm-25mm (0.2in-1in)
Armament One 73mm gun; two 7.62mm machine guns in bow; one 7.62mm machine gun co-axial with main armament; one launcher rail for AT-3 Sagger ATGW

WESTERN SPGs

With the increased mechanisation of armies in recent years, the towed artillery weapon has given way to the self-propelled gun or howitzer, which has the cross-country mobility to keep up with the armoured formations for which it provides fire support. Many of the newer self-propelled guns, such as the French 155mm GCT and the Italian 155mm Palmaria, have a semi-automatic or fully automatic loading system and attain a high rate of fire. In practice such a weapon would fire for a few minutes and then move off to a new position before the enemy could locate it and return fire. Self-propelled anti-tank guns are still in use, but most countries now prefer to deploy vehicles fitted with anti-tank guided weapons, which have a longer range.

Although SPGs (self-propelled guns) were in use before 1939, World War II acted as a significant impetus in the development of this hybrid type of weapon, half artillery and half tank. Since 1945 a wide range of SPGs has come into service, becoming an integral part of all armoured formations.

Self-propelled guns can be divided into two fairly distinct categories. The first is the tank destroyer or assault gun which is, in effect, a cheap tank. Usually lacking a turret and having less armoured protection than the conventional tank, the tank destroyer nevertheless has a large high-velocity gun and its main function – like that of the main battle tank – is to knock out opposing tanks. Relying for survival on manoeuvrability and careful defensive positioning, the tank destroyer lacks both overall power and flexibility when compared to the main battle tank – admittedly more expensive to produce – and only a few armies deploy this kind of SPG.

Few doubts, if any, are attached to the importance of the second type of SPG, the self-propelled artillery piece: a gun or howitzer mounted on a caterpillar-tracked vehicle provided with some form of armoured protection. A considerable degree of variation exists within this categorisation. Some SPGs have turrets while others rely on the vehicle moving to traverse the gun; a number are equipped with elaborate automatic loading devices and the level of protection afforded the crew is far from uniform.

The main role of self-propelled artillery is, as its name suggests, to act as highly mobile artillery, able to keep up with and support armoured units. Towed artillery, in contrast, must largely keep to hard road surfaces and thereby lacks this cross-country mobility; and, in addition, it cannot be deployed on the battlefield with the same speed as the SPG and is without any real means of crew protection. This latter point is of growing importance when the possibility of NBC (nuclear, biological and chemical) warfare is an ever-present danger, and consequently the more advanced SPGs are fitted with NBC protection.

During World War II Britain relied upon improvised SPGs – an artillery gun directly mounted on a tank chassis – so that the 25-pounder on the M3/M4 chassis became the Sexton and the 17-pounder anti-tank gun mounted on the Valentine tank became the Archer. After the war Britain relied upon American imports and it was not until 1958 that work began on a British gun which entered service in 1964 as the FV433 Abbot.

The Abbot was developed to utilise components from the FV432 APC series, notably the chassis, and is armed with a turret-mounted 105mm providing a traverse of 360 degrees and a maximum elevation to over 70 degrees. The gun has a maximum range of 17,000m (18,600yds) and can fire high explosive, smoke, target indicating, illuminating and HESH (high-explosive squash-head) ammunition. Supplementary armament comprises a 7.62mm light machine gun for anti-aircraft protection and smoke dischargers. Weighing less than 17 tonnes, the Abbot is a lightweight SPG, capable of a top road speed of 48km/h (30mph), and is fitted with a flotation screen enabling it to cross water obstacles. Infra-red driving lights and an NBC system are fitted. Although the Abbot serves with the Royal Artillery in Britain and Germany, the 105mm gun is now considered to be too small a calibre and has been superseded by SPGs of larger calibres supplied by the United States.

France produced a 105mm SP howitzer in the 1950s utilising the versatile AMX13 chassis, and this gun was supplemented by the larger 155mm Mk F3 in the early 1960s, similarly based around the AMX13. Although successfully exported to a number of Third World nations, both SPGs were relatively primitive weapons and in 1969 a new SPG was developed to replace them in French Army service.

Previous page: A US M110 self-propelled gun fires on Viet Cong positions at Fire Base Carol in South Vietnam. Above: The long-barrelled 155mm gun of the French GCT self-propelled gun is raised to maximum elevation.

Left: Britain's Abbot SPG was developed from the FV432 APC series and is armed with a 105mm gun. Right: The massive turret of the Italian Palmaria SPG houses a 155mm gun which can be fitted with either a manual or an automatic loading system. Below right: The West German Jagdpanzer Kanone features a 90mm anti-tank gun set in a sloping glacis plate. Bottom right: The Swedish VK-155's heavy and cumbersome automatic loading system is mounted directly above the turret.

The 155mm GCT is a heavy (42 tonnes loaded) and expensive weapon but has many advanced features. Mounted on an AMX30 tank chassis, the 155mm gun is enclosed in a turret providing a full 360-degree traverse. Maximum range with an HE projectile is 23,500m (25,700yds) and 30,000m (32,800yds) with a Brandt RAP (rocket-assisted projectile). An automatic loading system is employed which enables eight rounds a minute to be fired.

Following on from the success of her SP anti-tank guns during World War II, West Germany has produced a 90mm Jagdpanzer Kanone SP anti-tank gun. It has an effective battlefield range of 2000m (2200yds) and fires both HEAT (high-explosive anti-tank) and HESH ammunition, at a rate of 12 rounds per minute. The gun is set in a sloping glacis plate and an infra-red searchlight is mounted directly above the gun and moves in traverse and elevation with the gun. A total of 750 have been built for the West German Army and a further 80 ordered by Belgium which have been modernised to include an improved fire-control system incorporating a laser rangefinder.

Although the 155mm SP-70 is a joint West German/Italian/British venture, Germany is the project leader. The SPG's chassis is based on the Leopard 1 tank and the gun is a modification of the successful FH-70 towed howitzer, produced by these countries. The initial 12 prototypes were completed in 1976 and feature a high rate of fire with a maximum range of 30,000m (32,800yds). As with the FH-70 it can fire advanced Copperhead precision ammunition.

Italy already has a 155mm SPG, the OTO-Melara Palmaria which utilises the Italian OF-40 MBT chassis. The welded aluminium turret is positioned in the centre of the hull and the gun features either an automatic or manual loader.

The Swedish have produced their own SPG, the 155mm Bandkanon 1A or VK-155 which is a very heavy vehicle weighing some 53 tonnes and consequently suffering a penalty in mobility so that its maximum speed is only 28km/h (17mph). Sharing component parts with the S-Tank, the VK-155 entered service in 1963 and was an early example of an SPG having an automatic loading system. A magazine of 14 rounds is attached to the rear of the SPG's turret, providing it with a high rate of fire so that it can expend its entire magazine within a minute. The advantage of a high rate of fire for an SPG is that if it can loose off a barrage of shells at the enemy in a short space of time it can then move to a new position before

Top: Three US SPGs of the 1950s at the Aberdeen Proving Ground. From the left, they are a 155mm M44, a 155mm M53 and an 8in (203mm) M55. All three types have their rear spades dug in to absorb the recoil, and reveal the easy access to the firing compartment characteristic of American SPGs. Above: An early SPG, the US M7 Priest armed with a 105mm howitzer. Above right: A 155mm M40 'Long Tom' SPG being prepared for action by a British gun crew. Right: The 8in howitzer of the M55 in action.

its presence is detected.

The major manufacturer of self-propelled guns, both in numbers and types, has been the USA. During World War II, besides producing the small 105mm M7, America brought out the 155mm M40 'Long Tom' SPG which saw service in the closing stages of the war and was also used to considerable advantage in Korea. The 155mm M44 howitzer followed, based on the chassis of the M41 light tank which was also used on the 105mm M52 SPG. A larger calibre SPG was also considered necessary and in 1952 the prototype M55 SP howitzer was developed which mounted a 203mm (8in) gun, capable of lobbing an HE projectile 16,800m (18,500yds).

These 'early generation' SPGs were replaced in the 1960s by a new range of models. The M107 and the M110 were designed to use a common chassis, the M107 mounting a long-barrelled 175mm gun, the M110 being armed with the 203mm gun. Both types were notable for not employing an armoured fighting

Above: A British Army M107 being prepared for action. The rear spade is well dug in and the 175mm shell is about to be loaded into the breech. Right: An M107 is fired in Vietnam, where it was used for long-range interdiction work.

Right: A line of M110A1s of the US Army. They can be distinguished from the M110 by their longer barrels and muzzle brakes. Below: An M110 at maximum recoil during a bombardment of enemy positions in Vietnam.

compartment, so that only the driver is provided with protection. The M107 came into service in 1963 and its gun had, at that time, the very long HE range of 32,700m (35,750yds) and while the M110 lacked this range it could fire heavier and a wider variety of shell types, including nuclear rounds. Both models are now being phased out of service in favour of the 203mm M110A1/A2 which has a greater range over the M110, being able to fire RAP rounds to a maximum distance of 29,000m (31,700yds).

The 155mm M109 has become the most important of America's SPGs; over 3000 have been built and since its introduction in 1962 it has been exported worldwide. The M109 has proved a highly successful design: its turbo-charged diesel engine provides range with reliability and its gun – housed in a 360-degree traversing turret – in the later M109A1 variant has a range of 18,000m (19,700yds). Besides conventional HE, illuminating and smoke rounds, the M109 can fire tactical nuclear and chemical warfare ammunition. The M109 was widely used in Vietnam by both the Army and the Marine Corps, its low weight of 23 tonnes providing a useful degree of cross-country mobility in Vietnamese conditions. Like other modern SPGs in Western service the M109 is able to fire the advanced Copperhead laser-guided projectile.

While SPGs are more expensive than conventional towed artillery their obvious benefits in mobility and protection guarantee them an increasingly important place within the West's artillery forces. SPGs of the type of the European SP-70 will set the trend for the future.

Above: A photograph of the rear of a US M109 SPG taken during a break on manoeuvres, with M16 smallarms neatly piled alongside. Of interest is the ammunition rack at the rear of the turret ready for re-loading. Left: The turret of an M109. In contrast to a tank an SPG requires easy access for both crew and equipment, hence the number of open panels and hatches that are shown here.

Above and right: The SP-70 self-propelled howitzer. The SP-70 is fitted with an automatic ammunition re-supply mechanism, attached to the rear of the turret. Once the SPG's own supply of shells is exhausted, a speedy re-supply system is highly important if the gun is to operate effectively. The SP-70 utilises the gun of the tri-national FH-70 and the chassis of the German Leopard I tank, an exceptional combination of gun and chassis.

SOVIET SPGs

In recent years the Soviet Union has introduced into service a complete new range of tracked self-propelled artillery weapons. These include the 122mm self-propelled howitzer 2S1 (designated the M1974 by the United States), the 152mm self-propelled gun/howitzer 2S3 (M1973), the 152mm self-propelled gun 2S5, the 203mm self-propelled howitzer, and a massive 240mm self-propelled mortar (M1975). These all have greater cross-country mobility than their towed counterparts, and in the case of the 2S1 and 2S3 have complete armour protection for their crews. The ASU-85 self-propelled anti-tank gun is only employed by the Air Assault Divisions while the older ASU-57 has now been phased out of front-line service in the Soviet Union.

The idea of mounting a field artillery piece onto a caterpillar-tracked chassis was first developed in the 1920s. The great advantages of the SPG (self-propelled gun) over conventionally-towed artillery were, firstly, its far greater mobility which allowed it to accompany the forward deployment of armour and be at hand to bombard centres of enemy resistance with minimal delay, and secondly, the degree of protection offered to the SPG crew, enabling the artillerymen to operate in hostile environments, immune from smallarms fire and deadly shell splinters.

Since World War II the primacy of the tank as the single most important tactical element on the battlefield has not altered – despite the emergence of sophisticated anti-tank weapons – but it has been realised that to be successful the tank has to operate as part of an all-arms team including artillery. The advent of NBC (nuclear, biological and chemical) warfare and the ever-increasing mobility of modern armies has ensured that the SPG's importance has increased progressively since 1945, a fact acknowledged by the number of SPG types in service with the armies of both Nato and the Warsaw Pact.

Although an SPG looks similar to a tank it has never been expected to fulfil that role; far cheaper to produce than the heavy, well-armoured tank, the SPG is essentially a forward-operating artillery piece, although the lightweight assault gun is designed to operate as a cost-efficient tank-killer.

While the Soviet Union has not developed the range of SPGs available to Nato – and until recently relied extensively on conventional artillery – the first Russian SPGs emerged during World War II as a means of disrupting the German Panzer formations and supporting the Soviet T-34 offensives. The SU-122/152 series, which first saw combat in 1943, was an effective amalgam of the KV heavy tank chassis and a 122mm or 152mm gun. Towards the end of the war the KV chassis was replaced by the IS (Josef Stalin) chassis, and redesignated ISU-122/152. This SPG remained in Soviet Army service until the early 1970s.

The Soviet ability to utilise a tank chassis for SPG mounts ensured that the highly successful T-34 would be likewise employed. Armed first with a 76mm gun

Previous page: The business end of an ASU-85 assault gun. Top: ASU-85s parade through Moscow as part of a public demonstration of the Soviet Union's military strength. Above: An ISU-122 crosses the River Spree by pontoon bridge during the closing stages of World War II.

The 2S3 on parade (below) and in the field (left). Despite its formidable appearance the 2S3 – like other SPGs – has armour which is thick enough only to withstand smallarms and indirect artillery fire.

but soon uprated with an 85mm main armament, the SU-85 medium SPG was deployed as a direct support weapon for T-34 tank formations. Towards the end of the war the SU-85 was further uprated to take the 100mm M1944 gun and redesignated the SU-100; its high-velocity gun proved capable of dealing with the German Panther and Tiger tanks, being able to fire armour-piercing rounds, and an HE round up to 19,200m (21,000yds). In its capacity as the standard armoured support gun in the Soviet Army it was only replaced in front-line usage in 1957, and as an export success to the Warsaw Pact and pro-Soviet Arab nations it remains in service to the present day. Even as late as 1973 both Egypt and Syria were deploying SU-100 battalions with their armoured and infantry divisions, although obsolescent in face of the latest Israeli armour and anti-tank weapons.

Despite the Soviet Army's interest in SPGs, little development work in this field was carried out in the postwar years, the army apparently being content with their existing models. It was only in the 1970s that a new generation of SPGs came into service, which today form the basis of Soviet self-propelled artillery.

The 20-tonne 2S1 or M1974 was first observed at a military parade in Warsaw in July 1974, and consists of a modified 122mm D-30 howitzer mounted in a rotating turret upon an MT-LB chassis.

Above left: The heavy weight of the 2S3's main gun necessitates an asymmetrical arrangement of road wheels to provide extra support for the turret. Right: A unit of 2S3s maintains strict formation during a Soviet parade. Below: The high-velocity gun of the ASU-85.

Left: One of the most effective elements within the Soviet Union's artillery forces are the 2S1 SPGs, shown here taking a salute.

Embodying a number of features to be found on the PT-76 amphibious tank, it is fitted with seven rubber-clad road wheels with Christie-type torsion-bar suspension and two hydraulic dampers to provide a suspension system capable of traversing rough terrain. Two track widths are available, the broader of the two for movement across soft ground. The 240bhp diesel engine provides a maximum road speed of 60km/h (37mph) and affords some degree of frontal protection to the four-man crew. The driver sits forward at the front left-hand side of the SPG while the commander, gunner and loader are situated within the turret. In keeping with its PT-76 ancestry the 2S1 has an amphibious capability and its boat-shaped hull facilitates the crossing of water obstacles, which by means of track propulsion can be traversed at a rate of 4.5km/h (3mph).

The main armament has a fume extractor and is fitted with a double-baffle muzzle brake. Some 40 rounds of ammunition are carried and by employing a semi-automatic breech a rate-of-fire of five rounds per minute can be achieved. Ammunition types include HE and HEAT and it is reported that the 2S1 can also fire an RAP (rocket assisted projectile) to a range of 22km (14 miles).

The 2S3 or M1973 SP howitzer entered service in 1973-74 and like its smaller counterpart has a turret-mounted main gun. The chassis, however, is based on that which carries the SA-4 Ganef SAM (surface-

Above right: A side-view of a 2S1 reveals the clean lines of this SPG. Right: An ASU-57 is loaded onto a pallet ready for air transportation within the An-22 in the background. The ASU-57 can be para-dropped by utilising cluster parachutes. Below: ASU-85s drive off to an assembly point after having been landed by An-22 transport plane.

to-air missile) and as a consequence lacks the amphibious capability of the 2S1. Due to the considerable weight of the modified 152mm D-20 gun/howitzer the six road wheels are asymmetrically positioned in order to provide greater support for the turret. The 520bhp diesel engine enables the 24-tonne vehicle to travel at 50km/h (31mph). The crew configuration is similar to the 2S3 although an extra member can be carried. Ammunition capacity is 40 rounds and in addition to HE and APHE rounds it is thought that an HE/RAP round can be fired, capable of a highly impressive range of 24km (15 miles). In addition the 2S3 has a nuclear capability and can fire a 0.2 kiloton-yield nuclear round.

Both SPGs have standard NBC systems and although they possess only limited armour protection, their mobility and speed of deployment and redeployment (about one-fifth of the time of conventional guns) will considerably increase their chances of battlefield survival.

The 2S1 SPG is designed to replace the D-30 towed howitzer, and the 2S3 to replace the D-1 towed howitzer, as well as forming an SPG battalion within the Soviet tank division. Organised into six-gun batteries the Soviet SPGs operate either at battery or regimental level (three batteries) with front-line manoeuvre units. The high degree of low-level SPG/armoured unit integration contrasts strongly with Nato practice, in that each battalion commander has his own individual artillery support, although because of the overall rigidity of the Soviet artillery system other adjoining units within range cannot call upon another unit's artillery support. While the 2S1 SPGs are deployed at manoeuvre level, often in a direct-fire role, the 2S3s tend to be held back at

SOVIET SPGs

2S1
Crew 4
Weight 20,000kg (44,100lb)
Performance Maximum road speed 60km/h (37mph); range (road) 500km (310 miles)
Armament One 122mm D-30 howitzer, HE range 15,000m (16,700yds)

2S3
Crew 6
Weight 24,000kg (52,900lb)
Performance Maximum road speed 50km/h (31mph); range (road) 300km (186 miles)
Armament One 152mm D-20 howitzer, HE range 24,000m (26,250yds)

ASU-57
Crew 3
Weight 3450kg (7600lb)
Performance Maximum road speed 45km/h (28mph); range (road) 250km (155 miles)
Armament One 57mm Ch-51 anti-tank gun, HE range 6000m (6560yds)

ASU-85
Crew 4
Weight 14,000kg (30,900lb)
Performance Maximum road speed 44km/h (27mph); range (road) 260km (162 miles)
Armament One 85mm M-44 gun, HE range 15,300m (16,700yds)

Above: An ASU-57 depicted in winter conditions.

Below: These Western photographs of an ASU-57 reveal the rather primitive nature of its construction; and its small 57mm gun can have only limited use on today's battlefield.

regimental or divisional level in an indirect over-fire role.

In 1980 a new self-propelled heavy artillery piece was reported to be in service with front-line units; possessing a long-barrelled 203mm gun, it is thought to have a nuclear capability. Like the 2S1 and 2S3 it has been employed in Afghanistan, although details of the weapon are as yet unavailable.

The Soviet airborne forces have their own self-propelled guns, specifically designed for air-portability. The ASU-57 was introduced in 1957. Mounting a 57mm Ch-51 anti-tank gun, it was designed to be as weight effective as possible. The ASU-57 weighs under four tonnes and can be air-dropped from the Antonov An-12 transport aircraft, pallet-mounted with cluster parachutes and retro-rockets which are fired when the pallet hits the ground to provide an acceptably soft landing. The major weaknesses of the ASU-57 are that its 57mm gun is ineffective against modern armour and its own minimal armour makes it vulnerable even to infantry support weapons.

Altogether more effective is the ASU-85 which, while employing the PT-76 chassis, dispenses with its amphibious capability. The 85mm gun is effective against light armour and its own thicker armour provides more effective protection for paratroop units, which because of their airborne weight restrictions are woefully short of effective artillery support weapons. The increased weight of 14 tonnes means that the ASU-85 cannot be para-dropped but nevertheless is air-transportable. Each Soviet airborne division fields 18 ASU-85s and since their introduction in 1961 they have seen service in Prague in 1968 and more recently in Afghanistan.

Since the introduction of the SAU series in the early 1970s the Soviet Army's commitment to replacing its conventional towed artillery with SPGs has begun to gather momentum. Before the 1970s only a small proportion of the Soviet Union's vast artillery forces was mounted on tracked vehicles. In 1978, 17 per cent of Soviet artillery in front-line positions in Europe was self-propelled and by the early 1980s it was estimated that the proportion could be as high as 40 per cent. Clearly, in the light of these dramatic advances, the importance of Soviet self-propelled guns will steadily increase especially as they meet the modern requirement of mobility and firepower.

MODERN SUPPORT WEAPONS

Artillery fire support grew in importance on the land battlefield from the 17th century onwards, although actions were still settled by combat involving cavalry and infantry. Beginning in the late 18th century, technical improvements in the construction of guns and ammunition dramatically increased their power and range. By World War I the development of improved methods of reconnaissance and communications – in the shape of the aeroplane and the field telephone – enabled artillery fire support to become the decisive force on the battlefield.

The experience of World War I profoundly influenced modern support weapons design. One obvious example of this is the development of two specific types of gun for the anti-tank and anti-aircraft roles. Another requirement was increased mobility for artillery pieces. During the 1918 campaigns on the Western Front rapid advances through the enemy lines placed the batteries at the rear of the defences in danger of capture. The weight and bulk of the guns and the fact that they were mostly horse-drawn meant that a battery could not easily retreat to a new position and resume shelling in support of the defending troops. After 1918, therefore, gun carriages began to be designed with springs and pneumatic tyres to enable them to be towed by trucks. In more recent years artillery has been rendered even more mobile by the advent of the heavy transport helicopter, and also by the use of lightweight materials and by the ingenious design of the guns themselves. The US M102 105mm howitzer and the British 105mm Light Gun are representative of this trend.

Another solution to the problem of mobility is to mount the gun on the chassis of a truck or armoured fighting vehicle. Primitive self-propelled guns were used during World War I and since 1945 they have become the standard form of artillery in the armies on Nato's central front and with the Israeli Army. The self-propelled gun will probably come to replace the towed gun as the standard support weapon in most armies, despite its considerably greater cost.

Artillery has become even more effective on the modern battlefield with the development of improved organisation. In World War I it was found that the artillery took a long time in responding to calls for fire support. Consequently, during World War II the control of guns was decentralised to a lower level of the command structure as infantry battalions were given their own integral artillery units. The battalion commander no longer needed to contact a higher level of command in order to lay down a supporting barrage. The allocation of greater responsibility to the Forward Observation Officer in the selection of targets further increased flexibility. Technical developments since 1945 have accentuated this trend. The use of navigational aids, laser range-finders, fire-control computers and radios now means that a battery stands a good chance of laying the first salvo right on the target in under 100 seconds from receiving the request for fire support.

These improvements in the accuracy and speed of response of modern artillery have been combined with the increased destructiveness of modern shells

Above: The crew of an 84mm Carl Gustav rocket launcher prepare to fire the weapon. A Swedish design, the Carl Gustav delivers a powerful shaped charge, able to penetrate up to 400mm (16in) of armour plate at ranges of up to 450m (490yd).

to make the modern gun capable of doing the work that formerly might have been done by a whole battery. The application of mechanical loaders to guns such as the Swedish FH77 means that a salvo of three rounds could hit the target within 8-12 seconds; modern guns lay down a formidable weight of explosive, and the use of nuclear shells would be even more devastating. Thus has artillery developed the ability to dominate the modern land battlefield.

This type of indirect-fire artillery is less useful against tanks and armoured personnel carriers than against unprotected infantry. Specialised direct-fire weapons are used against 'hard' targets such as enemy tank formations, and these are now mostly guided missiles. Compared to the anti-tank guns used during World War II the modern missile is more manoeuvrable, but is handicapped by poor short-range performance. Anti-tank guided weapons also have a guidance system that normally demands some kind of operator assistance. The operator either directs the missile by means of signals sent down a wire to the missile itself, or by designating the target with some type of laser device. Obviously, an operator who comes under fire will not be able to pay much attention to guiding his missile onto its target, making the accuracy of these weapons in real combat open to question. These limitations mean that on the modern battlefield the primary anti-tank weapon is still likely to be the gun-armed tank.

Armoured fighting vehicles are still vulnerable to attack from the air. To counter the threat posed by the high-performance jet and the helicopter, the modern army is provided with a comprehensive array of air-defence weapons. These take two forms: the surface-to-air missile and the anti-aircraft gun. The mobility of these weapons is often increased by mounting them onto tank chassis, like the self-propelled gun. Their accuracy leaves something to be desired, despite the use of homing devices and radar direction (which are, however, susceptible to enemy jamming measures). The normal kill rate of air-defence weapons is between two and three per cent although anti-aircraft guns are more effective against helicopters. They do have a nuisance value, however, in that they are very distracting for the pilot, and may well put him off his aim.

As the technology of warfare becomes more complex, firepower and flexibility remain the cardinal virtues of any weapons system. This is as true of support weapons as it is of tanks or assault rifles, and as the accuracy, destructive power and mobility of the former increase, they will make an even more decisive impact on the modern battlefield.

Above: British paratroops alongside the Vigilant anti-tank missile launcher – one of the more effective direct fire weapons available to the infantryman.

Below: A US soldier lines up a Stinger surface-to-air missile during a field exercise in West Germany.

WESTERN FIELD ARTILLERY

The Western and Nato armies have fewer field artillery pieces than the Warsaw Pact forces, but compensate for this by adopting modern design features, particularly very advanced fire-control and target acquisition systems. Nato has standardised 105mm and 155mm as the basic calibres, and even non-Nato armies have fallen into line with this. Mobility, particularly by means of helicopters, is the principal asset of the 105mm weapons, while most 155mm howitzers are equipped with auxiliary propulsion to help them evade enemy counter-fire.

Amongst the many improvements made in the design and use of artillery during World War II were advances in mobility and versatility – the latter being the ability of a weapon to fulfil a dual role – the most notable example being the German 88mm anti-aircraft/anti-tank gun. One of the tactical lessons reaffirmed during the war was the importance of airborne or ground-based FOOs (forward observation officers) whose task was to control and coordinate artillery fire by watching the fall of shot and passing corrections to the gun line. The guns themselves ranged from close-support weapons to heavier guns and howitzers capable of longer range counter-battery fire designed to silence the opponent's own artillery.

Since World War II the importance of the field gun has in no way been reduced. Indeed, quite the reverse could be argued, especially with the introduction of shells capable of delivering a nuclear warhead, thereby providing artillery with a new-found strategic role. On the tactical side, conventional artillery played a central part in the Korean War, and even in guerrilla-style conflicts – such as the European involvements in Indochina – its effect has at times been crucial. The French defeat at Dien Bien Phu in 1954 is a good example of the battle-winning power of the field gun. The French positions were destroyed when they were subjected to a massive artillery bombardment which saw over 100,000 rounds pour into Dien Bien Phu. Unable to make an effective reply and with casualties

Page 127: A prototype of the multi-national 155mm FH-70 gun/howitzer on trial in 1974. Left: Although limited in range, the light-weight 105mm Model 56 pack howitzer is particularly effective in mobile warfare since it can be rapidly deployed into position by helicopter.

Below left: Viet Minh artillery pounds French positions at Dien Bien Phu. The Viet Minh victory owed much to the devastating effect of their well-sited artillery. Bottom left: French 105mm guns fire against Viet Minh enclaves in the hills along the Black River. Below: The devastating effects of the Copperhead CLGP (cannon-launched guided projectile). Copperhead has a range of 20,000m (21,900yds) and can be fired from a variety of artillery pieces including the 155mm M109A self-propelled howitzer, the 155mm M198 and the FH-70.

rising, the isolated French position was doomed. The Viet Minh had moved their guns into position by dragging them across the most inhospitable terrain, a feat which caught the French totally off guard and once again demonstrated the importance of mobility in warfare. Although the size and shape of field guns has not radically changed since 1945, they have continued to become more mobile and the latest artillery pieces have a greatly improved ability to be brought to firing positions with minimal delay.

Both the Soviet Union and the United States have devoted much finance and energy to improving their conventional artillery and several new types of gun have evolved. In the West some joint national ventures have been launched, and ammunition trends of more recent years have included the American M712 Copperhead CLGP (cannon-launched guided projectile). This relies on terminal guidance by an FOO who illuminates the target with a laser designator towards which the projectile (which is fin-stabilised and fitted with small wings) is guided by its electro-optical seeker. The round is available for users of the FH-70, M198 and M114 guns and its accuracy is extremely high in all weather conditions. Projects such as the Hughes Tank-Breaker and various extended-range guided projectiles show that development in this field is by no means static.

Artillery bombardments are a highly unnerving experience for those exposed to them and, for example, British artillery fire was a powerful factor in undermining Argentinian morale in the Falklands War of 1982. The high mobility of the helicopter-transportable Light Gun and the effective use of FOOs enabled the British forces to harass the Argentinian troops from frequently changing positions. The Argentinians were able to respond with 155mm artillery fire, but this was never effectively coordinated and consequently had little effect.

The 155mm FH-70 field howitzer is the result of a joint venture undertaken in the 1960s, initially between Great Britain and West Germany but later joined by Italy. In Britain the FH-70 was intended to replace the well-tried but ageing 5.5in medium guns (dating back to World War II) which were withdrawn from service in 1980. In concept the FH-70 was designed to accept a new series of ammunition and feature an APU (auxiliary power unit) which could be used to bring the gun swiftly into action under its own power without recourse to a tractor. In 1978 the first FH-70s entered service, with the UK ordering 71 for deployment in its 18-gun artillery regiments, West Germany 216 and Italy 164. The 6m (19ft 6in) barrel has a double-baffle muzzle brake and is provided with a semi-automatic wedge breech. The split trail carriage mounts an 1800cc VW APU which allows the FH-70 to be driven at speeds of 16km/h (10mph) and also powers the trail and main wheels. Only two minutes are needed to bring the howitzer into action.

The range of ammunition developed for the FH-70 includes an HE fragmentation shell which weighs

43.5kg (95.8lb), smoke and illumination rounds, and the Copperhead CLGP. A rocket-assisted projectile is also under consideration which would extend the weapon's range to over 30km (18.6 miles). In German and Italian service a digital display unit processes information from a fire-control computer, though the FH-70s of the Royal Artillery do not have this facility. In mid-1982 it was reported that Saudi Arabia had placed an order for the weapon.

In 1974 the 105mm Light Gun replaced the Italian 105mm OTO-Melara Model 56 pack howitzer which had served with the light regiments of the Royal Artillery. It extended the latter's range from 10.5km (6.5 miles) to 17km (10.5 miles) and has subsequently proved to be a very successful weapon. The gun's bow-shaped trail, which carries the turntable while on the move, and light weight of 1858kg (4096lb) allows it to be towed by a Land Rover or carried as an underslung load beneath a helicopter. Up to eight rounds per minute rate-of-fire can be achieved for the first minute, with a rate of three thereafter.

Ammunition types include a 16kg (35.2lb) HE round and it can accept smoke and illuminating rounds. The barrel is fitted with a detachable double-baffle muzzle brake, vertically sliding breech and hydro-pneumatic recoil system. The plus 70-degree elevation and 11-degree traverse – or 360 degrees on its turntable – contribute to its versatility. The Light Gun is in service with Brunei, Kenya, Malawi, Oman and the UAE, while Australia has it under order.

Top: Royal Artillery men crew the now obsolete 5.5in gun/howitzer. Developed in 1939, the 5.5 saw action in Africa from 1941 and remained in service until the late 1960s. Above: Italian troops prepare a 105mm OTO-Melara pack howitzer for action.

Below: The semi-automatic wedge breech mechanism of the FH-70 field howitzer. The FH-70 is crewed by eight men and is capable of a normal rate of fire of six rounds per minute. Currently under development is a self-propelled variant, designated the SP-70, which is expected to enter service in the late 1980s.

The US Army's 105mm M101 howitzer can date its origins to well before World War II but only after the end of the war did it receive its current designation. The gun's weight is just over 2 tonnes and elevation to plus 66 degrees and traverse to 46 degrees are provided. It is served by a crew of eight who can produce an initial rate of fire of eight rounds per minute which decreases steadily to the sustained rate of 100 rounds an hour. Maximum range is 11km (6.8 miles) and a wide range of ammunition is available, from anti-personnel, HE and anti-tank types to leaflet rounds. The weapon has been sold and exported to over 40 countries. The West German Rheinmetall FH-105(L) is a modified M101 with a longer single-baffle barrel. The FH-105(L) can deliver a 21kg (46.2lb) shell at a muzzle velocity of 600m/sec (1968ft/sec) to just over 14km (8.6 miles). It has a split trail and weighs 2.5 tonnes. The 105mm M102 howitzer was developed in the late 1950s and saw action in Vietnam. Today it provides the airborne and air mobile divisions of the US Army with a lightweight replacement for the M101. The gun has an aluminium carriage, with 360 degree traverse, and this has helped reduce the overall weight considerably to under 1.5 tonnes.

The United States Army's current 155mm howitzer is the M198. Development of this weapon began in the late 1960s and the first production examples entered service in 1978. The US Army has placed an order for 435 guns and the United States Marine Corps has ordered 282. The M198 will also be supplied to Australia, India, Greece, Holland, Pakistan, Saudi Arabia and Thailand. This split-trail weapon has a crew of ten and weighs 7 tonnes. A hydropneumatic recoil system is fitted and the 6m (19ft 6in) barrel has a double-baffle muzzle brake. It can accept standard types of ammunition as well as nuclear and Copperhead rounds. With a rocket-assisted projectile the range extends to 30km (18.6 miles) and representative of typical shell weight is the M692 HE round weighing 46.5kg (102.5lb) and containing 36 antipersonnel mines. An APU is under development.

Other types of American 155mm weapons include the M59 gun – known as the 'Long Tom' – (no longer

Top: A French artillery crew on manoeuvres with the American 105mm M101A1 howitzer. With a weight of 2.25 tonnes, the M101A1 is airportable by Chinook helicopter and can be deployed quickly and effectively in mountainous positions. Centre: American troops unpack ammunition for the 105mm M102 light howitzer. Designed to replace the M101, the M102 is lighter and has a slightly increased range.

Right: A British 105mm Light Gun in action against Argentinian positions near Port Stanley. Developed as a replacement for the Italian 105mm OTO-Melara pack howitzer the Light Gun saw extensive service in the Falklands.

Above right: A British 105mm Light Gun on NBC exercise in Germany. Above left: A French 155mm M1950 field howitzer at the Centre de l'Artillerie in Poitiers. The 9-tonne M1950 is the standard 155mm weapon deployed by the French Army with a range of 17,600m (19,250yds) and a firing rate of four to five rounds per minute. Below: The French 155mm TR which first appeared in June 1979 and known as Le Cannon de 155mm Tracte, is planned as the replacement for the M1950. The new piece has a considerably longer range and can fire the Brandt RAP (rocket assisted projectile) in addition to a 43.2kg (95.2lb) HE round, and illuminating and smoke projectiles. The TR is fitted with a double-baffle muzzle brake and can traverse through 65 degrees.

in US service but used by, among other nations, Austria, Japan, Italy and Turkey) and the M114 howitzer, which is retained by the US Army and also widely distributed among pro-Western nations. This latter weapon accepts the Copperhead round. The largest American field gun is the 203mm M115 howitzer. This fires a 92kg (202lb) HE round as well as nuclear shells. It weighs 14.5 tonnes, is served by a 14-man crew and has a range of 16.8km (10.4 miles).

Although France's motorised infantry divisions are currently equipped with 155mm M1950 towed howitzers which date from the late 1940s, plans are in hand to re-equip them with the 155mm TR towed gun, of which 130 have been ordered. Broadly similar to the FH-70 it has a barrel length of 6.2m (20ft 3in), a double-baffle muzzle brake and hydro-pneumatic recoil system. Served by an eight-man crew, the gun has a split-trail carriage and weighs some 10 tonnes. Barrel elevation is to plus 66 degrees and range is dependent upon ammunition type – with the Brandt rocket-assisted projectile, for example, it can extend to over 30km (18.6 miles). The gun is towed by a Renault TRM 6x6 tractor which also carries the 50 rounds of ammunition. A small APU is fitted along the lines of that of the FH-70. The towed gun can be brought into the firing line in five minutes, and an automatic rammer means that three rounds may be fired in the first 18 seconds, after which a rate-of-fire of six per minute is possible for the next two minutes.

The Italian OTO-Melara Model 56 pack howitzer has provided a 105mm weapon for numerous Western nations. Its great advantage is that it can be dismantled into 11 pack loads in three minutes which can then be transported by anything from mules to helicopters, and is thus ideally suited to mountainous terrain. Another advantage of the pack howitzer is its application in the anti-tank role where it is capable of penetrating up to 102mm (4in) of armour. It is served by a crew of seven and overall weight is only 1.29 tonnes. It can deliver a 21kg (46.2lb) projectile to over 10.5km (6.5 miles) and more than 2400 examples have been produced since the first guns appeared in 1957.

SOVIET FIELD ARTILLERY

The Russians have always called artillery 'The Queen of the Battlefield', employing it in massed 'artillery armies' during World War II to pulverise the opposition. Today this tactic is no longer feasible, but nevertheless Soviet artillery strength remains impressive, both in diversity of types and in sheer numbers. The general characteristic of Soviet guns is light weight for good performance in terms of range and shell weight. Their design is seldom revolutionary, but they are capable of standing up to rigorous service, and are used by many countries throughout the world.

Soviet military planners have always emphasised the importance of artillery and some 15 per cent of Soviet Army personnel are in artillery units. During the 1970s the Soviets placed great emphasis on the development of self-propelled guns, but towed artillery has not been neglected. It has been recognised, however, that it is no longer feasible to muster several dozen field pieces wheel-to-wheel, as was done during World War II, in order to swamp a target with fire, since such a concentration would rapidly be detected by modern sensors and attract countermeasures ranging from artillery to air and missile strikes. New tactics have evolved for the use of artillery as part of combined-arms operations, and at the same time a tendency has developed to move away from the deployment of shorter-ranging light calibre field pieces to reliance on heavier-calibre howitzers with the ability to command a greater area of the battlefield.

The backbone of Soviet artillery, from the time of the tsars until the 1960s, was the 76mm divisional gun in its various successive models. The last of these was the M1942 or ZIS-3, introduced in 1942 and produced in vast quantities during World War II and for some years thereafter. It was widely distributed throughout the Warsaw Pact armies and has been sold equally widely to Asian and African countries. By the 1980s it was only used in the Warsaw Pact as a reserve and militia weapon, and for training, but it could still be found in frontline service elsewhere. And no wonder: it is a thoroughly workmanlike and efficient piece of equipment, capable of serving perfectly adequately in any role where the calibre is suitable. The M1942's efficient design came from wartime rationalisation; its predecessor, the M1936, was efficient but heavy, and the M1942 was put together by using the M1936 barrel and marrying it to the carriage and recoil system which had been developed for a 57mm anti-tank gun. The result was a design which was 'right' from the start. Capable of firing a 6.2kg (13.7lb) high explosive (HE) shell to 13,300m (14,580yds), it was also provided with an armour-piercing (AP) shell and a hollow charge anti-tank shell, both of which were capable of dealing with most World War II tanks. As a result the M1942 was used interchangeably as a field or anti-tank gun.

In 1943-44 the Soviets developed an 85mm weapon to supplement the 76mm design. This went into service as the 85mm divisional gun D-44, though it did not reach the hands of troops until after the war had ended. A long-barrelled weapon on a split-trail carriage, it fired a 9.6kg (21.2lb) shell to 15,650m (17,115yds) and like its predecessors it was provided with AP and hollow charge shells, giving it a useful anti-tank capability. In this latter role it was frequently fitted with a large infra-red searchlight above the shield which, in conjunction with a special sighting telescope, allowed firing in darkness against tanks to a range of about 600m (650yds).

In postwar years the Soviets expanded their airborne forces and, in order to provide them with a useful dual-purpose gun, equipped them with the 85mm D-44. The only drawback was that this needed

Previous page: A 122mm M1938 (M30) howitzer is prepared for fire by Polish troops on Warsaw Pact manoeuvres. Above: The now obsolescent 76mm M1942 field gun, mainstay of Soviet divisional artillery in the 1940s.

Below: North Vietnamese troops blast away at US positions using an 85mm D-44 field gun. Lightweight and rugged the D-44 was highly regarded by NVA field units.

Right: The SD-44, photographed from the rear to show its two-cylinder 14bhp engine mounted on the trail. Steered by one of the gun crew the drive unit is capable of a top road speed of 25km/h (15mph) and provides the SD-44 with a useful degree of cross-country mobility.

the presence of a towing vehicle, and so the ingenious Russian designers produced the first 'auxiliary propulsion' weapon to enter military service, a fashion which has since spread. The SD-44, as it was known, had a two-cylinder 14bhp petrol engine mounted on the trail so as to drive the gun wheels by a conventional prop-shaft and differential. The trail ends were supported by a single wheel which could be steered by the driver/gunner who sat on the trail ahead of the motor. On a hard road the device could reach a speed of 25km/h, and the engine could be de-clutched to allow the gun to be towed in the normal way if a tractor was available. The SD-44 stayed in service with Soviet airborne divisions until the late 1970s, at which time they were restructured as rifle divisions and their heavy support removed.

By the 1980s the D-44 was used only for training in the Soviet Army, but was still believed to be in frontline service with other members of the Warsaw Pact; it has also been widely spread throughout the Third World countries.

At a higher calibre, a 100mm field gun, the BS-3, was put into service in 1944. The barrel came from an existing naval gun and the carriage followed conventional practice, being a split trail with two dual wheels and a shield. As a field gun it fired a 15.6kg (34.4lb) shell to 21,000m (23,000yds) and with a very long barrel it could deliver a 16kg (35lb) AP shell at 1000m (1100yds) per second to pierce 185mm (7.3in) of armour at 1000m (1100yds) range, a very potent performance. It was later provided with a discarding sabot shot which had a velocity of 1415m (4641ft) per second and could pierce 200mm (7.9in) at the same range, and a hollow charge shell capable of holing 380mm (15in) of plate at any range.

Unfortunately, like most of the heavy anti-tank guns of its day, the 100mm gun was a heavy and cumbersome weapon to manhandle around the battlefield, weighing 3650kg (8050lb). It was gradually replaced in Soviet service by a specialised 85mm

Right: Leftist Muslim militia fire on Christian positions north of Beirut with a Soviet-made 122mm D-74 field gun. The D-74 can fire a 27kg (60lb) shell to a maximum range of 24,000m (26,250 yds).

Below: An Egyptian 122mm A-19 (M1931/37) field gun dug-in prior to an Israeli assault in the Sinai during the Six-Day War of 1967. Developed as a long-range counter-battery weapon during World War II, the A-19's reliability in action and its long range will ensure its continuing use in many Third World armies. Bottom: Deployed in line a battery of 122mm M1938 (M-30) howitzers awaits orders to fire.

anti-tank gun and later by a smoothbore 100mm anti-tank gun which weighed some 650kg (1430lb) less. It is still widely distributed outside the Soviet Union.

During the heyday of the 76mm field guns, the next step up was the 122mm medium artillery howitzer. The M1938 went into service in 1939 and is still present throughout the Warsaw Pact and Third World armies in vast numbers. Like the 76mm M1942 it was a sound design which can still give good service, firing a 21.8kg (48lb) shell to a range of 11,800m (12,900yds). However, its range was considered insufficient for modern warfare, and in the late 1940s it was supplemented by the 122mm gun D-74, a long-barrelled weapon capable of firing a 27kg (60lb) shell to 24,000m (26,250yds). This appeared to be satisfactory in delivering a heavier shell to a longer range, but at the same time another design bureau had developed a totally new weapon, the 130mm M-46 gun, and in comparative trials this latter proved to be even more efficient. As a result, comparatively few examples of the D-74 were made and most of them were exported, equipping among others the Egyptian, Chinese and Cuban Armies.

The M-46 was once again an existing naval gun adapted to field use, by fitting it to a two-wheeled split trail carriage. Firing a 33.4kg (73.5lb) shell to 27,000m (29,500yds), it also had an AP shell capable of piercing 230mm (9.1in) of armour at 1000m (1100yds) range, giving the weapon a very useful anti-tank performance. Its capabilities were first seen by Western agencies in Vietnam, where it was deployed from 1971 onwards. The gun rapidly acquired a reputation for phenomenal accuracy as well as destructive ability, and a remarkable quantity of US airpower was devoted to an unsuccessful effort to take the M-46 out of the war. The Egyptians used the M-46 in a counter-battery role during the 1973 Arab-Israeli

Above left: A Soviet gun crew fires a 130mm M-46 gun. Possessing a range of 27,000m (29,500 yds) the M-46 has proved highly successful in long-range artillery duels, notably in Vietnam where NVA artillery was able to outrange most US field guns with ease. Above: While Soviet armour advances in the background a battery of 122mm D-30 howitzers are prepared for action. With its wheels in the up position the D-30 has a quick 360 degree traverse, and in this picture the tow attachment can be seen below the muzzle brake.

Left: The D-30 in action – Lebanese Muslim troops pound Christian positions around Beirut from a mountain stronghold. Designed to fire both anti-tank and conventional high-explosive rounds the D-30 is a highly versatile weapon.

War with a considerable degree of success. It has been widely exported to other Asian and African armies, as well as being the standard field piece of the Warsaw Pact forces.

While the 122mm M1938 howitzer was an efficient weapon, it lacked flexibility for modern warfare, and in the early 1960s a replacement, the D-30 122mm howitzer, appeared. This was a thoroughly modern design, which was apparently based on a Skoda weapon developed for the German Army in 1943–44. Instead of the conventional split trail and two wheels, it has a platform with three outrigger legs which can be spread, and the wheels lifted from the ground, so as to allow the howitzer to be traversed and fired through the full 360 degree circle. This permits rapid employment against targets in any direction, very desirable in a war of movement. The D-30 is not towed by the trail, but by the muzzle. It fires the same 21.8kg (48lb) shell as the M1938 but to a maximum range of 15,400m (16,840yds); it also has a rocket-assisted shell which boosts the range to 21,000m (23,000yds), though with some degradation of accuracy. The howitzer has seen service in the 1973 Middle East war, and in Nigeria, Angola and Ethiopia.

The next step in traditional calibres was to 152mm, a class which had existed in Russian service since the 1870s. The current 152mm weapon in service is the D-20 gun-howitzer, mounted on the usual sort of split-trail carriage. This fires a 43.5kg (96lb) shell to a range of 17,400m (19,000yds) as standard, and is also reputed to be provided with a rocket-assisted shell which reaches to 37,000m (40,000yds) and a 0.2 kiloton nuclear shell. It is in service throughout the Warsaw Pact and in Yugoslavia, is still being manufactured in China, and has probably been supplied to various Asian and African armies.

The largest Soviet field piece is the 180mm S-23 gun, another design which began life as a naval gun

SOVIET FIELD GUNS

D-30
Crew 7
Calibre 122mm
Weight Firing 3150kg (6944.5lb)
Range 15,400m (16,840yds), rocket assisted 21,000m (22,966yds)
Ammunition HE 21.8kg (48.1lb) HEAT 14.1kg (31.1lb) and 21.6kg (47.6lb), smoke, chemical and illuminating

M-46
Crew 9
Calibre 130mm
Weight Firing 8450kg (18,629lb)
Range 27,000m (29,528yds)
Ammunition HE 33.4kg (73.5lb), AP 33.6kg (74.1lb)

S-23
Crew 15
Calibre 180mm
Weight Travelling 20,400kg (44,974lb)
Range HE 30,500m (33,355yds), rocket assisted 44,000m (48,120yds)
Ammunition HE 84kg (185lb), AP, 0.2kt nuclear, G-572 concrete piercing

and was then grafted on to a field carriage. The carriage in this case is a large split trail design with twin wheels, the trail ends being carried on a two-wheeled limber when being towed. The gun barrel is drawn back in its recoil system and strapped to the trail while travelling, in order to reduce the vibration on the unsupported barrel. The S-23 fires an 84kg (185lb) shell to a maximum range of 30,500m (33,400yds). It is also provided with a concrete-piercing shell for attacking fortifications, a rocket-assisted shell which extends the maximum range to 44,000m (48,000yds), and a 0.2 kiloton nuclear shell. Employed by the Egyptian Army during the Yom Kippur War of 1973 the S-23 demonstrated its ability for accurate long-range fire. In Soviet service the S-23 is employed by the Heavy Artillery Brigade of the Artillery Division; it is not in general Warsaw Pact use, and outside the Soviet Union this 180mm gun can be found with the Indian, Syrian and Egyptian Armies.

Above: A Soviet 152mm D-20 gun-howitzer in its firing position, with wheels splayed outwards. Assigned at divisional level the D-30 acts in a general support capacity. Below: The largest of the Soviet Army's field guns, 180mm S-23s photographed here in Red Square, 1968.

ANTI-TANK WEAPONS

The threat of the main battle tank continues to preoccupy the minds of the military planners, most particularly those of the Nato armies when they consider the imbalance in armour strength between Nato and the Warsaw Pact. This has resulted in a continuous development of anti-armour weapons, ranging from the recoilless gun to shoulder-fired rockets and missiles of varying degrees of sophistication. The struggle is, of course, self-perpetuating; as soon as the weapons designer produces a new warhead, the tank designer appears with a thicker and stronger tank. In 1944-5 the conventional anti-tank gun could overcome the tank only at the expense of being impractically big and heavy; whether today's weapons will go the same way remains to be seen.

The radical improvements in tank design that took place during World War II led, not unnaturally, to counter-developments in the field of anti-tank weaponry. The high-velocity anti-tank gun came into general service as the surest means of dealing with the massively armoured tanks that dominated the battlefields of World War II. And yet that war saw the first developments in a new class of anti-tank weapon – the anti-tank missile. Initially primitive hand-held devices, they were to be transformed into a range of electronically-sophisticated guided weapons capable of extraordinary degrees of accuracy; and while the gun has remained a potent anti-tank weapon, the most dramatic advances made in this field since 1945 have come about through progress in ATGWs (anti-tank guided weapons).

The anti-tank gun's main advantage over the ATGW is its ability to fire all three types of anti-tank ammunition – AP (armour-piercing), HESH (high-explosive squash-head) and HEAT (high-explosive anti-tank) – and while the gun itself is an expensive item the rounds of ammunition are relatively cheap and plentiful, especially when compared to the projectile costs of the more advanced ATGWs.

Ranged against the anti-tank gun are its weight and size which make it an unwieldy object on the battlefield, rendering it and its relatively exposed crew highly vulnerable to small-unit infantry support weapons as well as to standard counter-battery fire.

Following on from the experience of World War II the Soviet Union continues to place faith in anti-tank guns and today has two principal types in service. The 85mm D-44 gun is a conventional World War II design – and has seen extensive use on the battlefields of the Middle East – while the larger 100mm T-12 is a smooth-bore weapon capable of firing the latest high-velocity APFSDS (armour-piercing fin-stabilised discarding-sabot) rounds. With a muzzle velocity of 1500m/sec (4920ft/sec), the T-12 is effective against thick armour up to a range of 1500m

Page 139: A Hughes 500 MD Defender armed with a mast-mounted sight (above the rotor blade) fires a TOW anti-tank missile. The mast sight allows the helicopter to engage enemy armour at low-level while remaining hidden by undulations in the ground. Left: Based on the M113 APC (armoured personnel carrier) an Emerson ITV (improved TOW vehicle) lets loose its TOW missile. Below left: Soldiers of the British Army set up a 120mm Wombat recoilless rifle.

(1640yd). On the other side of the Iron Curtain, West Germany, Belgium and Switzerland deploy self-propelled 'tank destroyers', armed with 90mm quick-firing high-velocity guns based on the success of the German Jagdpanther of World War II. The other Nato armies, while not employing the conventional high-velocity anti-tank gun, ensure that their field artillery batteries can be used in a direct-fire anti-armour role, which, for example, in the case of the US 105mm M101 howitzer enables it to penetrate 100mm (4in) of armour plate up to a range of 1500m (1640yd).

At the infantry level, the lightweight recoilless rifle has been brought into service. By allowing some of the propellant gases to be vented directly through the breech on firing, it eliminates the need for the heavy recoil-absorbing buffer recuperator system which makes conventional artillery so unwieldy. The British Army chose the 120mm Wombat recoilless rifle, capable of firing a HESH round to a distance of 1000m (1100yd), which can be mounted on an infantry-manned two-wheel trailer, a long wheel-base Land Rover or the FV432 armoured personnel carrier. Apart from a somewhat limited armour-piercing capability, the recoilless rifle suffers from the problem of a large back-blast which gives away the firer's position – a potentially fatal situation for the advanced and supposedly well-hidden recoilless-rifle crew. Possessing no significant advantages over the modern anti-tank missile, the recoilless rifle is now being phased out in favour of the ATGW.

The simplest anti-tank missiles on offer today are shoulder-launched weapons, termed LAWs (light anti-tank weapons), which utilise the lightweight HEAT warhead. The LAW operates on either the rocket-launched or recoilless system. The US Bazooka was one of the first and best known of the rocket-launched systems; although even modern versions have a range limited to around 200m (220yd) its lightness makes it attractive to the over-burdened infantryman. The Soviet RPG-7 combines a light

Above: An Iraqi soldier prepares to fire his Soviet-designed RPG-7 rocket launcher, supported by comrades armed with AK assault rifles.

Right: A soldier of the Royal Irish Rangers aims his 66mm LAW which features a disposable launch tube. Below: While needing a crew of two men the 84mm Carl Gustav LAW has a more powerful warhead than the lightweight 66mm LAW.

launcher with a relatively large warhead by adopting a two-stage launch system: the first is a small charge which propels the rocket clear of the launcher after which the main rocket is fired to carry the warhead to the target. The RPG-7 has its drawbacks, however, most especially a lack of flight accuracy caused by the mid-air ignition knocking the moving – and potentially unstable – missile off course.

Of the recoilless LAWs the Carl Gustav FV550 launcher has an improved range of 700m (765 yd) but its complicated sights increase its weight to 15kg (33lb) and two men are needed to operate the weapon. In an attempt to solve the weight problem, the French Strim rocket launcher uses a novel system of firing the projectile from within a discardable package tube which absorbs most of the recoil. This facility has made it possible for the launcher to be built of lightweight materials.

Whether the LAW incorporates the rocket or recoilless system has begun to seem less relevant, however, for improvements in MBT (main battle tank) protection – especially the Chobham compound armour – have rendered sophisticated tanks virtually immune to light anti-tank weapons. While LAWs can be usefully employed against softer targets such as MICVs (mechanised infantry combat vehicles) and APCs (armoured personnel carriers), for use against MBTs they need to be replaced by more complex and heavier ATGWs which can send a powerful warhead to ranges considerably in excess of the anti-tank gun's shell.

In the short history of the ATGW three generations of missiles have been developed. The first – including the Soviet Sagger, the British Vigilant and the French SS 11 – are all wire-guided missiles using 'command to line-of-sight guidance' which entails an operator using a joy-stick to relay signals down the command wire and by simple optical means to guide the missile directly onto the target. The second generation is more complex, though by using a semi-automatic guidance system the operator's task is in fact made simpler; his function is to track the enemy vehicle by keeping his sights aligned to the target for the time it takes the self-correcting infra-red homing, wire-guided missile to hit it. The Franco-German HOT (high-subsonic optically-tracked tube-launched) missile and the US TOW (tube-launched optically-tracked wire-guided) missile are the two most important semi-automatic ATGWs in the Nato armoury and, while they both have excellent maximum ranges of 4000m (4375yd) and 3750m (4050yd) respectively, their bulk forces them to be taken away from the foot soldier and to be set on vehicle mountings. Less heavy is the Franco-German Milan (*missile d'infanterie léger anti-char*) which is just man-portable and has equipped a number of European armies as their main infantry anti-tank weapon.

At long ranges the ATGW missile can take some time to hit its target – 17.5 seconds at 3000m (3280yd) for the West German Mamba, for example – and under stressful battlefield conditions it has been found that the longer the flight duration the less accurate the missile: it is not easy to hold a target in your sights for 17 seconds when you are under fire. Allowing the operator to use a 'fire-and-forget' system improves matters considerably, as has been suggested by the semi-active Hellfire missile. The Hellfire can be fired from either a helicopter or from behind cover and guides itself to the target by picking up the laser beams

Above: Well-sited in a prepared position a Vigilante anti-tank missile team lines up a potential target.

Like its TOW counterpart, the HOT anti-tank missile can either be launched from a helicopter – such as the West German MBB Bö105 (right) – or from a ground vehicle (above).

Above: The technologically advanced AH-64 Apache attack helicopter is armed with 16 Hellfire anti-tank missiles. Left: The Soviet Army's wire-guided anti-tank missile, the AT-3 Sagger, is highly portable. Right: German and French anti-tank helicopters are now fitted with TOW or HOT missiles as, for example, the four TOW tubes on this Bö105.

reflected off the target by a well-concealed, forward laser operator.

The more advanced systems need to be mounted on tracked vehicles which because of their size and vulnerability on the battlefield need to be so well armoured that they begin to resemble the armoured vehicles they are set against. Somewhat ironically many armoured warfare theorists regard this as a logical consequence of the fact that the tank itself is the best anti-tank weapon. Armed with a large-calibre high-velocity main gun, the modern MBT is superior to the ATGW in the short-range engagements that make up most modern tank versus anti-tank encounters. The introduction of compound armour on the latest MBTs has lessened the effect of the chemical-action HEAT and HESH rounds and as a result the armour-piercing kinetic energy round has regained much of its former importance. The latest high-velocity tank guns fire a sub-calibre (discarding-sabot) solid shot (either tungsten carbide or depleted uranium) held on course by fin-stabilisers, so that on contact the hardened shot has a very high specific gravity which allows it to pierce a tank's armour.

While the kinetic energy APFSDS round is the most important anti-armour weapon of the MBT, HEAT and HESH chemical-action projectiles have been most successfully deployed in the latest attack helicopters. Although exaggerated claims were made for the anti-tank attack helicopter, both HOT and TOW missiles have been fitted to Nato helicopters thereby providing them with the opportunity of attacking large armoured formations at relatively safe distances. In many ways the highly manoeuvrable attack helicopter is the ideal weapons platform for the anti-tank missile. The advanced US AH-64 Apache anti-tank helicopter is fitted with electronic devices to suppress hostile ground fire and numerous other built-in survival features, while at the same time being capable of launching 16 laser-guided Hellfire missiles according to a flexible tactical plan that ensures a very high number of direct hits. On the Soviet side a similar interest in anti-tank helicopters has been shown. The introduction of the Mi-24 Hind attack helicopter in the early 1970s caught many Western observers by surprise. A powerful if cumbersome aircraft, the Hind's first role in a European war would be to provide an extra cutting edge to the otherwise blunt instrument of the Red Army; and with the introduction of the Hind-F – able to carry six Spiral laser-guided fire-and-forget anti-tank missiles and a 30mm armour-piercing gun – the Soviets have an advanced and very specialised anti-tank weapon.

In its different way the fixed-wing aircraft can be highly effective against tanks. Of the many munitions that can be used against armoured vehicles, one of the most potent is the cluster bomb, a collection of hollow-charge mini-bombs that break away from the parent weapon to form a scattered spread of destruction over a wide area. Of the latest improved conventional munitions the most cost-effective has been the anti-tank mine, which in the case of the German Medusa and Pandora series can be sown quickly and efficiently from helicopters. Although such mines are unlikely to knock out an MBT they can cause sufficient damage to both crew and vehicle to halt its progress.

As military planners in both the West and the East regard the tank as the most important single element in the land forces' armoury, the efforts made to counter this threat will be correspondingly great. The dramatic improvements in conventional weapons accuracy that began to take effect in the late 1960s and 1970s have now been fully extended to anti-tank weapons. In the future it seems likely that the range and application of anti-tank weapons will continue to expand so that the MBT will be faced by a concerted all-arms offensive ranging from the simple hand-held LAW, through land-based ATGWs to the latest electronic guidance systems for shells and 'smart' bombs.

Top: The Hellfire ATGW. Above: The Hellfire sight is slaved to the helmet-mounted optical unit of the gunner. Below: An M48 in flames, victim of a Hellfire.

ANTI-AIRCRAFT ARTILLERY

In spite of the development of air-defence missiles, there are still large numbers of anti-aircraft guns in use today and, surprisingly, the number is increasing. This is almost entirely due to advances in electronic technology which permit very accurate fire control to be compressed into small packages which can be carried on small-calibre weapons. Another reason is the improvement in ammunition technology which permits shells of between 25mm and 35mm calibre to be extremely efficient against modern aircraft. The high rate of fire of these small-calibre automatic guns, coupled with modern, sophisticated micro-computer control systems, gives a very high hit probability on targets at ranges below 3km (2 miles).

The end of World War II found air-defence artillery at its zenith; every combatant nation had lavished research and manufacturing ability on the provision of large numbers of guns of every calibre from 20mm to 150mm, together with radar and fire-control systems, ammunition and all other relevant equipment. However, the war had also generated the weapon which was eventually to reduce this vast army of guns to a tenth of its wartime strength: the surface-to-air missile (SAM). Under development in Germany, Britain and the United States, SAMs were expected soon to replace the gun as the leading anti-aircraft weapon.

Programmes to update air-defence artillery continued in the postwar period, however, especially as the running down of development budgets slowed missile programmes. In 1944 the US Army had initiated a totally new concept when they specified a 75mm gun with automatic feed and with the radar and fire-control integral to the gun mounting. The idea of an automatic feed had been made possible by the development of the electronic proximity fuze: the major obstacle to rapid fire had lain in the need to set a time fuze before loading every shell, to ensure that the shell would burst within range of its target, but the proximity fuze detected the presence of the target by radio and detonated the shell when it was in the most advantageous position, without requiring any setting before firing.

The US 75mm M51 Skysweeper did not appear until the mid-1950s, since development encountered more obstacles than had been anticipated. The gun was fed by twin rotating magazines and could achieve a rate of fire of 45 rounds per minute (rpm). Alongside the gun, on the mounting, were a tracking radar, a fire-control computer and an optical tracker, so that the entire engagement of the target was in the hands of the gun's crew and they had no need for external sources of data. Skysweeper is still in service with the armies of Greece, Turkey and Japan, but it was replaced in US service in the early 1960s by the Hawk missile.

In general, by the 1960s heavy AA guns of 75mm calibre or larger had been made obsolete by SAMs. The British experimented with two heavier guns for high-altitude air defence in the 1950s, but they also eventually adopted a SAM, the Bloodhound. Some Soviet heavy AA guns were used in Vietnam around Hanoi and by Arab states in the Middle East during the 1960s and 1970s, but with no great success.

The lighter guns, however, still had a useful service life. Light AA guns, generally of 37mm or 40mm, were common during World War II – everyone had heard of the famous Bofors gun – but the increasing speed of jet aircraft had left them behind by 1950. Hand-operated, they were incapable of swinging fast enough to engage a ground-attack aircraft moving at 800-950 km/h (500-600 mph), so power operation was adopted, first as an 'add-on' to existing guns and then in a purpose-built design when Bofors produced their improved L/70 model in 1951. Such designs were normally used in conjunction with a central fire-control radar and, combined with improvements in shells (giving a greater destructive radius), in the cartridge (giving better velocity and increasing range to 3000m – 9850 feet) and in the rate of fire (now 300 rpm), gave the light anti-aircraft gun a new lease of life. Some examples of other light anti-aircraft guns

Previous page: A West German Gepard in open countryside during the Operation Lionheart exercise of 1984. Top: An M163 of the US 59th Air Defense Artillery providing protection for the airfield at Baumholder. Above: Inspection of the linked ready-use ammunition of the Vulcan; a variety of ammunition is available for the Vulcan, including armour-piercing.

Left: The M998 Sergeant York prototype. The radar was developed from that in use on the F-16, the guns are Bofors 40mm L/70 and the chassis came from an M48A5 tank. Despite the M998's many advanced technical features, during trials it was found to be no more effective than the Vulcan system and was scrapped because of its vast cost.

produced since the war are the Oerlikon twin 35mm GDF-002 and the Soviet 57mm S-60, both of which are still in service.

During World War II there were many attempts to achieve a higher mobility for AA guns by putting them onto the chassis of tanks or transport vehicles, thus enabling them to accompany armoured columns. The US Army replaced its wartime M16 and M19 self-propelled AA guns with the M42, which entered service in 1952. This was armed with two 40mm M2A1 cannon (which has proved effective against the armour of PT76 light tanks) and equipped with a lead computing sight. A range-only radar was experimentally installed but no version of this was made. The M42 served with the US Army until 1969 and remains in service with the armies of other nations today.

The advances in electronic and optical techniques which took place in the 1960s and 1970s made it possible to incorporate micro-computers and similar small electronic devices into practical sights, and also allowed radar sets to be greatly reduced in size until they could be easily accommodated in relatively small gun mountings. One of the first vehicles to take advantage of these developments was the West German Army's Gepard, introduced in 1971. Based on the hull of the Leopard 1 tank, its turret carries surveillance radar, tracking radar, a fire-control computer, and two Oerlikon 35mm belt-fed automatic cannon. The surveillance radar scans all round the vehicle and can detect targets up to 15km (10 miles) away; as soon as it detects a target the crew is alerted and the bearing, range and elevation angle are displayed, together with an indication of whether the aircraft is friend or foe. If it is hostile, then the information is automatically transferred to the track-

Above left: The Bofors 40mm L/70 gun, successor to the famous Bofors m/36; this example has the Bofors Optronic Fire-control Instrument (BOFI) enabling night operation. Above: An M42 of the Lebanese Army in Beirut.

Below: One of the newest types of towed AA guns, the Artemis 30mm system, is made by Hellenic Arms Industries using Mauser guns.

ing radar on the front of the turret, which begins to track the target, releasing the surveillance radar to search for fresh targets.

Information from the tracking radar is now fed to the fire-control computer, which is also informed of wind speed and direction, temperature, barometric pressure and other factors. These are all applied to the target data, and gun elevation and bearing are calculated and passed to the power-operating function which swings the turret and elevates the guns. At this point the commander can, if he wishes, 'take over' the target in his optical sight, or he can leave the tracking radar in control. Whichever is in operation, when the target is within range of the guns the gunner is given a visual signal to open fire. The twin guns combine to produce a rate of fire of 1100 rpm and firing usually begins when the target is about 4km (2.5 miles) away. The level of accuracy is extremely high – so high, indeed, that when the Belgian Army took Gepard into use they complained that it was too accurate and had a compensating factor entered into the computer which distributes the fire around the notional point of aim to introduce a spread of shot and give a greater danger area. While undoubtedly effective, Gepard is extremely complex and expensive. It is also large – 47 tonnes – and is therefore likely to be confined to major armoured formations.

When the US Army came to replace its M42, it incorporated a radar into the M163 Vulcan. Deliveries of this vehicle began in 1968 and it is still produced for export; it has seen action with the Israeli Army and reportedly shot down a Syrian Sukhoi Su-7 during the 1982 invasion of the Lebanon. The M163 is an M113 armoured personnel carrier fitted with a one-man turret, incorporating a lead computing sight and a range-only radar. Capable of firing either 1000 or 3000 rpm, to a range of 1600m (5200 feet), the Vulcan proved unsatisfactory in service and work on a replacement began in 1977.

The US Army introduced the Sergeant York DIVADS (divisional air-defence system) in 1983. Now discontinued because it was so costly, this was a modified M48 tank chassis with a turret carrying twin 40mm L/70 Bofors guns, search and tracking radars, optical sight with laser rangefinder, and fire-control computer. The engagement of a target was almost entirely automated; the radar detected and informed the fire-control computer which interrogated the target and computed the firing data, the open-fire range, type of ammunition, burst length and interval, and took into account target manoeuvres, weather conditions and barrel wear. When all the calculating was done the gunner merely gripped his controls and the guns opened fire at the computer's command. Two types of ammunition were stored in magazines and could be fed to the guns as necessary; one shell was impact-fuzed, while the other was proximity-fuzed and filled with high explosive and 640 pellets of tungsten alloy so as to give the optimum fragmentation pattern for target damage.

Smaller equipments than this are required for lesser duties, and here the smaller calibre cannons come into play, 20mm and 25mm being the preferred types. The French Panhard M3 VDA is a good example of this class of weapon.

The M3 VDA (*véhicule de défense antiaérienne*) is a four-wheeled Panhard armoured personnel carrier with a turret devised by the Electronique Serge Dassault company. The turret contains a search radar, an

Top left: A Czech M53/59 self-propelled AA system in action. It is armed with two 30mm M53 cannon. Above: The Panhard M3 VDA was first shown at the 1975 Satory Exhibition; 95 per cent of its mechanical components are identical with those of the AML armoured car. Left: An AMX30 self-propelled AA system advancing through swampy terrain; the twin 30mm guns can fire up to 650 rounds per minute.

Far left, top: ZSU-57s on parade for May Day, 1960. First seen in public in 1957, the ZSU-57 uses a heavily modified T54 chassis. Far left, centre: Considered by some to be one of the finest self-propelled AA systems in the world, the ZSU-23-4 suffers from a tendency to shoot while traversing after firing.

Far left, below: The Soviet 37mm M1939 AA gun was based on the design of the Bofors 40mm m/36. It has seen action in Vietnam and during the Arab-Israeli Wars, and is shown here in the Lebanon.

electro-optical sight, and two 20mm cannon. The search radar is small enough to be carried under a man's arm when removed from the turret, but powerful enough to detect targets at ranges up to 8km (5 miles). It is also capable of tracking a target while still scanning the surrounding area for further targets. Once a target is detected the radar sounds an alarm, interrogates, and if hostile feeds the target range and elevation to the optical sight. At the same time it automatically slews the turret to face the oncoming target, and the gunner then begins to scan vertically with his sight until he acquires the target optically. The fire-control computer, fed with data from the radar, has calculated lead angles and other firing data, and this displaces the visual aiming mark. The gunner now lays his sight on the approaching target, and as soon as it is within engagement range a light flashes in his field of view and electric restraints on firing are switched off. He then opens fire manually while tracking the target, the guns firing at either 200 or 1000 rpm from each barrel.

20mm shells are now thought to be marginal in their effect against modern armoured aircraft, and most designers have begun to move up to 25mm or 30mm weapons which produce a lethal effect but which are still convenient in size and cost. One of the newest systems to be offered is the West German Wildcat, a wheeled gun carrier with a turret mounting two 30mm Mauser cannon. Wildcat has a search radar with a range of 18km (11 miles) and this displays targets on a screen in front of the gunner. It also directs the electro-optical sight unit in the desired direction. A television camera detects the approaching target, locks on to it and begins tracking; a laser rangefinder then measures the range and continues to measure it, so establishing the target's speed. The computer calculates the firing angles and also evaluates the target data and determines the number of shots to be fired in each burst to obtain optimum kill probability. As soon as the target is within range, a signal is given and the gunner can open fire. The reaction time of the system, from detecting an approaching target to being ready to open fire, is about six seconds. The Mauser 30mm gun is also employed in some of the latest towed equipment, such as the Greek-made Artemis and the Italian Breda Twin 30.

It is interesting to compare the Soviet equivalents of these systems. The Soviet Army has used self-propelled air-defence guns since the early 1950s, long before the Western armies began thinking about them, and there are two presently in service, the ZSU-57 and the ZSU-23. The ZSU-57 uses a chassis based on T54 tank components, carrying a large turret armed with two 57mm guns. These guns are the Soviet development of a design which originated in Germany in 1942 but which the Germans never completed. They are fed from four-round clips and have a rate of fire of no more than 70 rpm for each

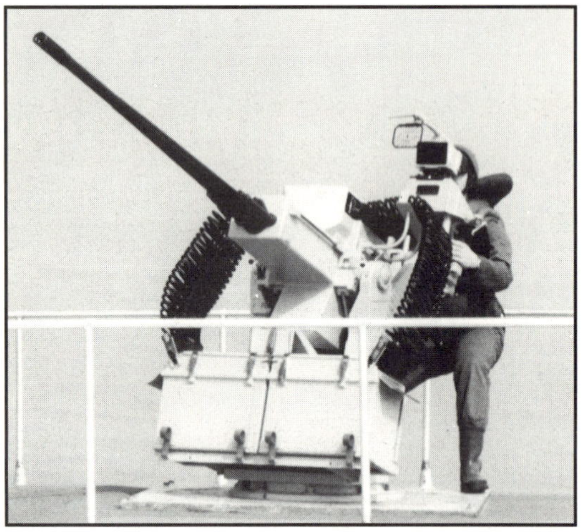

Above: A Gepard opens fire with its 35mm guns; these can fire either single shots, in bursts or continuously. Left: The Oerlikon 25mm GBM-AO1 naval mount; naval AA guns disappeared from ships with the advent of the SAM but they are now being retrofitted to deal with the anti-ship missile and low-flying aircraft. Centre left: An Oerlikon 35mm GDF-001 of the Swiss Air Force. Below left: Reloading of the automatic replenishers of the Oerlikon 35mm AA gun; rapid fire is of great importance against jet aircraft. Below right: A West German Wildcat which uses the Mauser 30mm AA gun mounted onto the chassis of a Transportpanzer 1. Right: The latest Anglo-French AA gun, the Chieftain-Sabre, uses 30mm AA guns and the AMX30's Oeil Vert surveillance radar.

barrel. The sighting system is simply an optical computing sight, the computing being done by a clockwork mechanism which is wound and primed with estimates of the target's range and speed. Not surprisingly, the ZSU-57 is not considered much of a threat by Western air forces, and it appears to have been phased out of Soviet first line service.

The ZSU-23 is more effective; it is based on a chassis derived from the PT76 amphibious tank, carrying a turret armed with four 23mm automatic cannon firing at a rate of about 900rpm per barrel. There is a dual-purpose radar which can detect targets out to 20km (12 miles); these are displayed on a screen and when one is selected, the radar is switched to tracking mode, though it can only do this once the target is within 8km (5 miles) range. Data is then fed to a computer which calculates firing data and presents it to the gunner, who then points his guns in accordance with this information and opens fire; alternatively he can use an optical sight to acquire the target and open fire. The ZSU-23 is more effective than the ZSU-57, and has replaced it in most Warsaw Pact formations, but it still falls short of current Western techniques.

In spite of the missile, the air-defence gun is still a useful weapon. One reason, as already explained, is that the gun mounting and sighting systems have been vastly improved due to modern electronic technology; another reason is simply that a 30mm explosive shell will bring down an aircraft or helicopter just as surely as a shoulder-fired missile and at about one thousandth of the cost. An air-defence gun system may be expensive to buy, but its running costs are infinitely cheaper than those of a missile system, and where the targets can be expected to come thick and fast, this makes a great deal of difference.

NATO SAMs

In common with several other postwar weapons systems, the SAM (surface-to-air missile) was developed from the research studies of the scientists of Nazi Germany. The unwieldy first generation of SAMs has since been replaced or supplemented by more mobile examples of modern, miniaturised technology, such as the Rapier and Stinger. These types of weapons are considered an important part of modern armed forces, and those produced by members of the Nato alliance are in use with countries throughout the world.

The ever increasing capabilities and numbers of Soviet attack aircraft has necessitated the provision of adequate AA (anti-aircraft) defences for Nato's land forces and strategically important installations in western Europe. The performance of the latest ground-attack aircraft is such that the traditional AA gun has, for the most part, become inadequate to counter the threat. Increasingly, therefore, reliance is being placed on the SAM (surface-to-air missile) to provide the necessary cover for effective operations in the face of the Warsaw Pact's numerically superior air forces.

The need for a flexible AA response has given rise to three distinct families of SAMs; man-portable, mobile (either towed or vehicle-mounted) and static systems. The first of these is intended to provide the humble infantryman with his own AA cover and, as such, uses simple aiming systems and (with some exceptions) infra-red guidance. The mobile systems usually comprise a launch unit and separate guidance and control equipment. They are used to provide cover for army formations, the defence of important strategic locations, and, on fully mobile mountings, they act as an integral part of armoured formations. The last category, the static system, is a permanent installation sited to provide cover for airfields, production centres and urban areas. Such systems, together with those which are mobile, usually rely on radar as the guidance and control medium.

Currently, France, West Germany, Italy, Sweden, the UK and the USA are producing SAMs. Taking these in turn, the major French system is the Thomson-CSF/Matra Crotale. Developed originally for a South African requirement, the R440 missile is an all-weather weapon optimised for use against low-flying aircraft. The system as a whole is made up of a launcher vehicle and a target AC (acquisition and coordination) unit, both mounted on a four-wheeled armoured chassis. The launch vehicle carries four missiles and is equipped with a J-band tracking and guidance radar. The AC unit has an E-band radar for target surveillance, identification and designation and carries a data-link system to pass information to the launcher. Maximum detection range is in the region of 18km (11 miles) and up to 12 targets can be handled by the AC unit at any one time. Currently, Crotale is in service with the French Air Force who use it for airfield defence, and the system has been widely exported, especially in the Middle East.

France, in collaboration with West Germany, also produces the Roland system. Like Crotale, Roland is optimised for use against low-flying aircraft and is mainly carried on fully-mobile tracked launchers based on the French AMX chassis or that of the German Marder. To date, there have been three variants of the basic weapon, Roland I and II and US Roland. Roland I is a clear-weather system which uses a launcher-mounted Siemens/Thomson-CSF surveillance radar for target acquisition and an optical sight for aiming. Roland II is an all-weather system which uses a Thomson-CSF tracking radar in addition to the Siemens/Thomson-CSF surveillance unit. As a back-up to the radar system, the optical sight is retained in this second model. US Roland is generally similar to Roland II and is mounted on a launch

Previous page: A British Bloodhound Mk2 long-range SAM blasts skywards. The Bloodhound has a maximum range of 80km (50 miles). Above: A French Crotale area-defence SAM is fired from its launcher. The missile's radar dish is visible, centred in the quad launcher.

Below: Mounted on a modified M109 tracked chassis a US Roland missile is fired on a test range. Despite the fact that the Roland is a highly effective short-range missile, it was abandoned by the US armed forces for cost reasons.

vehicle based on the American M109 chassis. In all cases, guidance commands are transmitted via a radio link and in the case of Roland II /US Roland, initial acquisition of the missile in the sighting system after launch is by means of infra-red. Once acquired, tracking is passed to the radar which simultaneously follows both the target and the missile.

Currently, the French armed forces operate both Roland I and II whilst those of Germany use only the second model. US Roland was intended to be the US Army's major SAM for the 1980s but cost overruns led to the whole programme being scrapped with only 27 launchers and 595 missiles delivered.

Italy's Spada system is based around the air-launched Aspide missile and is intended as an all-weather, point/area defence weapon for small strategic areas such as harbours and airfields. A Spada battery comprises between two and four static, six-round launchers, a search and IFF (identification-friend or foe) interrogation unit using an E/F-band radar, a tracking unit using a G/H-band radar and a control unit using an I-band radar for missile guidance. Operationally, these elements are grouped into two sections known as the detection centre (an overall command post and the search unit) and the firing section (the tracking unit, the control unit and the launchers). The detection centre handles target identification and allocation whilst the firing section performs target acquisition, tracking and missile guidance. Currently, this rather complicated system is only in service with the Italian Air Force.

Although outside Nato, Sweden's RBS-70 SAM is operated by Norway within the alliance in addition to Sweden itself and the Irish Republic. Developed by Bofors, the RBS-70 is a man-portable system comprising a launch tube, sighting unit and firing stand, each element constituting an individual man-load. The complete weapon can be assembled in roughly 30 seconds and guidance is provided by laser designation. Operation is simplicity itself, it only being necessary to keep the sighting unit aligned with the target during the engagement. A laser designator is slaved to the sight and the missile homes in on the target by 'riding' the beam of laser light.

The UK is a major producer in the SAM field and currently has operational the Blowpipe man-portable weapon, the Rapier mobile system and the static Bloodhound. Developed by Shorts, Blowpipe comprises a launch tube containing the missile and a clip-on aiming unit containing a command radio transmitter, an 'auto-gathering' unit and a monocular sight. To fire the missile, the operator acquires the target in the sight's cross wires, releases a safety catch and fires. Once in flight, the missile is automatically 'gathered' so that it remains centred on the sight and the operator guides the weapon by the movement of a thumb-operated joystick. To engage another target, he simply unclips the empty launch tube and replaces it with a fresh unit. In addition to the UK, Blowpipe is used by ten other countries including Argentina. This led to the bizarre situation of both sides in the Falklands War using the weapon against each other, with the British claiming nine and the Argentinians one Blowpipe 'kill'.

The Rapier system was designed originally as a missile replacement for the British Army's faithful 40mm Bofors anti-aircraft gun. In its basic form, Rapier comprises a four-round launcher with a built-in surveillance radar, a generator unit and an optical

Above: Perched on a Land Rover two Swedish soldiers fire a Bofors RBS-70 SAM. The weapon has a range of 5km (3 miles) and utilises a laser-designation system to guide the missile onto the target. Right: The multiple Blowpipe system in operation; the soldier on the right is guiding the missile onto its target while his companion is overseeing the operation through binoculars.

Right: A British soldier on exercises with a Blowpipe SAM. Although missiles such as the Blowpipe are designated 'man-portable' they are clearly cumbersome weapons to transport around the battlefield. Below: The Blowpipe missiles at the moment of launch. The Blowpipe is a useful infantry defence weapon and was used with some degree of success in the Falklands conflict.

tracker. Operationally, a target is acquired by the radar set and if deemed hostile, passed to the tracker operator. He then follows it through his optical sight and when it comes within range, a missile is automatically fired. The operator then continues to track the target and an in-built computer processes the missile's deviation from the 'line of sight' and generates guidance commands accordingly.

This basic Rapier has been combined with a fully-mobile tracked launcher based on the M548 cargo chassis. The towed system has been further developed into an all-weather weapon by the addition of a trailer-mounted DN181 Blindfire radar set which provides for a fully automatic engagement sequence. In addition to the British services, Rapier is in service with or ordered for 11 other countries including Switzerland and the USA. Rapier has been used in action on the Falklands (claiming 20 'kills') and by Iran in the Gulf War.

The remaining British SAM is the venerable Bloodhound Mk2 which is now used solely for airfield defence in the UK. Typically, a Bloodhound battery comprises of four missiles and static launchers, a target illuminating radar and a launch control post.

America has developed a wide range of SAMs which currently comprise the man-portable Redeye

Right: Men of the Royal Artillery prepare the tracker element of a Rapier SAM for operation. The launch system can be seen in the background with its four Rapier missiles ready for firing. The Rapier is a far more complex SAM than the simple infantry Blowpipe and is accordingly able to engage aerial targets in a wide variety of tactical situations.

Left: The Rapier SAM in action. Below: The Rapier system installed on the M548 cargo chassis, which extends tactical mobility and provides ground troops with improved air defence.

Left: A line of RAF Bloodhound Mk2 SAMs of No. 85 Squadron, defending a British airfield. Although its great range is a useful asset, the Bloodhound's ability to deal with the latest combat aircraft remains in question.

siles were replaced in the Soviet Union by more advanced types though they are still used by the Warsaw Pact and in other countries which deploy Soviet weapons.

The SA-4 Ganef marked a considerable improvement on its predecessors in having a greatly increased range (although at a reduced speed of Mach 2.5), while at the same time being carried by a highly mobile launch vehicle. Consisting of two missiles mounted on a tracked launcher, the SA-4 system has the ability to advance with the swiftest mobile forces and yet provide effective air-defence protection to a range of 70km (43 miles). Introduced in 1964 the SA-4 has not been used in combat; although a number were stationed in Egypt during 1970-72 they were withdrawn prior to the Yom Kippur War in 1973. An updated version was brought into operation in 1974 with improved capabilities at low altitudes and it is thought that this missile may have its own terminal radar-homing system.

The SA-5 Gammon can hardly be considered a conventional SAM, for with a launch weight of 10,000kg (22,050lb) and a range of 250km (155 miles) it forms part of the Soviet national air-defence system.

When the SA-6 Gainful was first observed in a Red Square parade in 1967 few Western observers had any idea of the capabilities of this self-propelled, triple-launcher missile system. In 1973, Egyptian SA-6s destroyed a considerable number of Israeli aircraft during the first few hours of the Yom Kippur War and it rapidly gained a fearsome reputation. This was largely unjustified, however, for despite its manoeuvrability, sophisticated terminal radar guidance and low-level capability, by the end of the fighting Israeli

Above: A captured SA-2 of the Egyptian armed forces is inspected by curious Israeli troops following the campaign in the Sinai during the Six-Day War of 1967. Right: SA-2s have been distributed throughout the Warsaw Pact and here a detachment of Polish troops answers an alert during an exercise held in 1981.

pilots had come to terms with the SA-6, having mastered it through a combination of evasive action techniques and by the timely release of 'chaff' clouds which disturbed the missile's guidance system. The SA-6's weaknesses were revealed further during the Israeli invasion of the Lebanon in 1982 when Israeli defence suppression aircraft wiped-out the Syrian SA-6 air-defence system with contemptuous ease.

Although the SA-6 had proved a useful and mobile weapon the Soviet authorities saw its limitations and in 1974 brought out the SA-8 Gecko. Considerably more advanced than the SA-6, the SA-8 is considered to have the necessary acceleration to hit fast-moving aircraft at low altitudes. The launch vehicle allows four missiles to be carried, which can be fired in pairs at separate targets. The tracking radar is carried on the launch vehicle (especially designed for amphibious operations) and provides the SA-8 missile with a maximum range of up to 15km (9 miles). Not having been used in combat, little is known of its true capabilities, however, although its radar guidance system is thought to be supplemented by an infra-red terminal homing system.

Distinct from the radio-command guidance systems of most SAMs are the SA-7 Grail and SA-9 Gaskin missiles which use a very simple infra-red homing system designed to lock onto the exhausts of passing aircraft. The SA-7 can be carried by individual infantrymen and the latest models are capable of shooting down slow-moving aircraft at ranges of up to 5.6km (3.5 miles). More advanced, the SA-9 is quad-mounted on scout cars and is equipped with a larger warhead. The great advantage of these portable weapons is that they allow units as small as an infantry section to have their own SAM defence, and their cheapness and simplicity of operation make them available to insurgent groups the world over. But as with other simple weapon systems, the SA-7 and SA-9 are easily countered and few modern combat aircraft would experience real problems from such weapons, although helicopters and slow-moving counter-insurgency aircraft remain vulnerable.

The latest generation of Soviet SAMs are the

Below: East German SA-4s are prepared for action.
Opposite top: SA-4s carry out manoeuvres in Eastern Europe. In the advent of war with Nato the SA-4 would play an important role in attempting to fend off Western strike aircraft from disrupting the advance of the Soviet Army.
Opposite bottom: The portable and cheap SA-7 infra-red SAM is given a firing demonstration by a Soviet soldier.

SA-10 and SA-11 which have been designed to counter low-flying strike aircraft such as the US F-111 and the European Tornado as well as US cruise missiles. The SA-10 is a long-range weapon of high manoeuvrability combined with exceptional acceleration, and it is thought to pose a real threat to Nato strike forces. The SA-11 is a shorter range SAM and is fitted on triple or quadruple launchers. From 1978 it began to replace the ageing SA-6 batteries within the Soviet Union.

The strength of the Soviet SAM system lies not in any one particular missile – for any missile can be mastered over time – but in the fact that Soviet military planners see the whole SAM programme as a continually evolving one, each new development providing a springboard for further improvement. Because of this the Soviet Union has been able to develop missiles as advanced as the SA-10 and SA-11 that look set to provide Nato air forces with their greatest challenge yet.

Overleaf top: SA-8 launchers (left) stand side-by-side with truck-borne SA-2s. Soviet air defence theory emphasises integrating all aspects of the SAM system. Overleaf bottom: An SA-9 launcher is shown here with only its two outer missile boxes in position on the launcher rails.

THE KEY SOVIET SAMs

SA-2 Guideline
Length 10.6m (35ft)
Launch weight 2300kg (5065lb)
Guidance Radio command
Fuel 1st stage, solid; 2nd stage, liquid
Maximum speed Mach 3
Maximum range 35km (21.7 miles) – later models 50km (31 miles)
Warhead 130kg (287lb) high explosive

SA-3b Goa
Length 6.1m (20 ft)
Launch weight 950kg (2094lb)
Guidance Radio command
Terminal homing Semi-active radar
Fuel Two-stage solid fuel
Maximum speed Mach 3.5
Maximum range 18.3km (11.4 miles)
Warhead 60kg (132lb) high explosive

SA-4 Ganef
Length 8.8m (28ft 10in)
Launch weight 2500kg (5512lb)
Guidance Radio command
Terminal homing Semi-active radar
Fuel Four solid-boost motors, plus ramjet sustainer
Maximum speed Mach 2.5
Maximum range 70km (43.5 miles)
Warhead High explosive; weight unknown

SA-6 Gainful
Length 6.2m (20ft 4in)
Launch weight 550kg (1213lb)
Guidance Radio command
Terminal homing Semi-active radar
Fuel Two-stage solid boost motor, plus ramjet
Maximum speed Mach 2.8
Maximum range 37km (23 miles)
Warhead 80kg (176lb); half of which is high explosive

SA-7 Grail
Length 1.3m (4ft 5in)
Launch weight 9.2kg (20.3lb)
Guidance Infra-red homing
Fuel Two-stage solid fuel
Maximum speed Mach 1.5
Maximum range 3.5km (2.2 miles) – later models 5.6km (3.5 miles)
Warhead 1.8kg (4lb)

SA-8 Gecko
Length 3.2m (10ft 6in)
Launch weight 200kg (441lb)
Guidance Radio command
Terminal homing Infra-red (or possibly semi-active radar)
Fuel Dual-thrust solid fuel
Maximum speed Mach 2
Maximum range 15km (9.3 miles)
Warhead 40kg (88lb) high explosive

SA-9 Gaskin
Length 1.37m (4ft 6in)
Launch weight 50kg (110lb)
Guidance Infra-red homing
Fuel Solid
Maximum speed Mach 2
Maximum range 7km (4.4 miles)
Warhead 5kg (11lb)

MODERN INFANTRY WEAPONS

Some theorists tell us that future wars will be fought by men sitting in deep bunkers and pressing buttons, using nuclear weapons, and 'Star Wars' space technology. However, the fact remains that the only way a country knows it has been defeated is when the inhabitants look out of their front doors and see an enemy soldier standing at the end of the street with a rifle in his hands. An occupied country is not defeated until it is seen to be defeated, and the occupation has to be done by foot soldiers. 'Tanks conquer, artillery supports, infantry occupy' is an old truism which is likely to remain true as long as wars are fought.

Because the infantry soldier operates on foot most of the time, there is a tendency to write him off as an unintelligent plodder; in the old days it used to be said that provided a man could master the use of rifle, bayonet and shovel he was a good infantryman. Those days have gone for ever; the infantryman of today has to be the most versatile man on the battlefield, for the range of equipment he is likely to be called upon to operate is far wider than that found in any other combat arm. A tank crew might be conversant with two or three different tanks, and they will not be all that different; an artilleryman may have to know how to operate three or four different guns or howitzers. The infantryman, on the other hand, has to master his personal weapon – the rifle – to the point where it is an extension of his arm and responds automatically to his will; but he must also be familiar with a sub-machine gun, one or two machine guns, two or three entirely different types of anti-tank weapon, two or three anti-aircraft weapons, half a dozen grenades, and two or three mortars. He must then duplicate most of this in learning enough about foreign weapons to be able to put to use anything he picks up on the battlefield. He must then learn to operate infra-red viewers, image-intensifying sights, seismic detectors, radios, computers for calculating firing data for certain weapons; he must learn to drive armoured infantry fighting vehicles and armoured personnel carriers and to operate the special weapons which they carry; he has to be able to read a map accurately, camouflage himself, dig defensive works, cook, find his way by the stars, and live off the country. He must be able to marshal a helicopter into its landing zone and marshal it out again, leap in and out of landing craft ... we could go on, but the point is surely made.

The infantryman's basic weapons are described in the pages that follow. They all have to be small and light enough for him to carry into action, and although some modern weapons stretch his strength and endurance to the limit, the tendency today is to miniaturise. Thirty years ago the standard infantry rifle weighed close to 5kg (11lb) and fired a bullet of about 7.5mm calibre. Today the weight is down to below 4kg (9lb) and the bullet is of less than 6mm calibre, so that the infantryman can carry more ammunition for his lighter weapon. But as a result of this reduction in weight and calibre the effective range of the personal weapon has dropped from about 600m (660 yds) to about 300m (330yds). This is not as great a deficiency as it sounds, however: comprehensive trials and examination of actual combat experiences have shown that the rifle is rarely used at ranges greater than 300m. Indeed, few soldiers can even see a target at greater distances than

Above: Using his Bergen (rucksack) for cover, a British soldier guards his position with a 7.62mm GPMG, the British version of the Belgian-designed FN MAG. This machine gun has been widely exported throughout the non-communist world.

Above: Armed with SLRs and a GPMG, part of an Infantry section races over an obstacle course. These two weapons have been the mainstay of the British Army at section level during the 1960s and 1970s, and performed with great credit during the Falklands War.

Below: In a rather theatrical pose in front of the Union flag, four British soldiers display their 7.62mm SLRs. This rifle replaced the old .303in Lee Enfield during the 1950s, and although very popular with the troops, it began to be replaced by the IW rifle during the mid-1980s.

this, since today's soldier is dressed in camouflage clothing and merges into his background at long range.

Argument is most vocal at present on the matter of machine guns for the infantry squad; several countries, notably the United States, Britain and the USSR, have opted for small-calibre light machine guns as the squad weapon. Other countries – France and Germany, for example – have set their faces against this, claiming that the light machine gun needs longer range and more striking power than the 5.56mm bullet can provide, and that the 7.62mm cartridge is the lower limit for machine-gun use.

Another weapon which is beginning to appear in larger numbers is the dedicated grenade launcher; by 'dedicated' we mean a weapon designed purely for grenade launching and not an adaptor attached to a rifle. This was first introduced by the US Army with their 40mm M79, and a few similar weapons have appeared since. But at the same time the Americans developed an automatic grenade launcher, feeding grenades from a belt and capable of pouring out a rapid stream of destructive projectiles to a range of about 250m (270yds). This met with little success and only a handful were ever made. But in the late 1970s the Soviets appeared with a similar weapon, one which has been seen on ground mounts, on the turrets of infantry fighting vehicles, and mounted in helicopters. Since then there has been a resurgence of interest in the automatic launcher in the West. The American model has been revived and put into production, and we expect to see one or two more designs appearing before long.

The sub-machine gun continues to puzzle most analysts; one moment it seems consigned to the scrap-heap, the next it is back in favour. Several new designs were introduced in 1983-84, but the lightness and handiness of the assault rifle has threatened completely to supplant the sub-machine gun in military service. However, short-barrelled versions of assault rifles, offered as sub-machine gun substitutes, have not proved as satisfactory as had been hoped. This is principally because they still fire the 5.56mm rifle cartridge, a cartridge which demands a specific length of barrel in which to develop its performance. When the barrel is cut back below that critical length, muzzle flash and blast rise to uncomfortable levels and the accuracy of the weapon drops in proportion. It seems that the sub-machine gun may yet find a battlefield application, but at the moment it is mostly used in counter-terrorist operations.

This introduction does not seek to give the impression that the armies of the world are at odds over their basic equipment. Where dissension exists it is no bad thing, since it makes the soldiers aware of other available options, and keeps them sufficiently critical to be able to argue with weapons designers when new ideas are proposed. The soldier's aim is quite simple; he wants the best weapon that will do the job. The trouble often lies in getting his political masters to define exactly what the job is.

MORTARS

Although the concept of the high-angle mortar dates from the 15th century, it is the Stokes mortar of 1915 which is the ancestor of today's weapons. With a few exceptions, today's mortar is a drop-fired smoothbore gun firing a finned bomb, and advances in design have turned the mortar from a cheap area weapon to a more expensive but much more accurate device capable of precise supporting fire. In several armies 120mm mortars are replacing light artillery, particularly where air-portability is of prime importance. The gun-mortar, developed in France, is also becoming more common on armoured vehicles where it can not only provide close support using a high-explosive bomb, but also be effective in an anti-tank role using an armour-piercing projectile.

Previous page: British paras prepare to fire an 81mm light mortar. Above: The FV432 APC mortar variant mounting an L16A1 mortar. The mortar has a full 360 degree traverse and is provided with 160 rounds of HE ammunition. Right: Light mortars can be broken down into three main elements and carried by infantry over difficult terrain.

The infantry mortar is the personal support artillery of the battalion commander, directly under his control and possessing the ability to respond very quickly to calls for fire. The near-universal pattern for mortar construction is of a weapon divided into three elements: a smoothbore barrel, resting on a baseplate which transmits the recoil thrust directly to the ground, and supported at the desired angle of elevation by a bipod. The bomb is a streamlined teardrop with fins at the tail end to give stability. The tail unit carries a shotgun-type cartridge of powder in its centre, and auxiliary charges of smokeless powder are clipped around the fins. When the bomb is loaded, by dropping it into the muzzle tail-first, it slides down the barrel until a fixed firing pin at the bottom strikes the central ignition cartridge, whereupon the auxiliary charges are fired and the resulting explosion lifts the bomb out and discharges it towards its target.

The mortar, by definition, fires only above 45 degree elevation, so that the bomb describes a high, arching trajectory which carries it over obstacles and delivers the bomb very steeply on to the target. This angle of approach means that the bomb is almost vertical when it detonates, and thus can distribute lethal blast and fragments in all directions; in this, it differs from an artillery shell which strikes the ground at a much shallower angle and is more directional in its effects.

Given these basic features, it is not surprising that most mortars have a resemblance to each other, even in calibre. Almost every Western army uses an 81mm infantry mortar, while the armies of the Warsaw Pact use an 82mm model; the difference between the two is negligible, though in general the mortars in use in the West are of more modern and efficient design, are more accurate and have something like 40 per cent more range than do the Soviet designs.

An example of technological superiority is seen in the British 81mm L16 mortar, which replaced the wartime 3-inch model. One of the problems with mortars is that the bomb must be smaller than the interior of the barrel, so that it can slide down, displacing air as it does so. But when the cartridge explodes, the gas which should be lifting the bomb is able to leak away through the 'windage' between bomb and barrel, so reducing the theoretical performance. Wartime bombs were made of cast iron and had a smoothly machined belt around the point of largest diameter; some had grooves machined in the belt so as to cause turbulence when the gases tried to pass and thus make some sort of a seal. The 81mm design, however, has the bomb made of special graphite iron and carefully machined all over. Around the waist is a groove into which a plastic sealing ring is snapped. This remains in the groove during loading, and allows the bomb to be dropped into the barrel. But when the explosion gases rise alongside the bomb, the shaped

Below left: A French Hotchkiss-Brandt MO-120-RT-61 120mm mortar in the firing position. The barrel is rifled and fires a range of ammunition including the 18.7kg (41.2lb) PRPA rocket-assisted bomb which has a range of 13,050m (14,750 yds). Below right: Troops training with an Austrian 120mm M-60 mortar.

Right: An Israeli 160mm self-propelled mortar carrier. Israel has modified a number of AFVs to meet its own requirements and the 160mm Soltam mortar has been mounted on both the M7 and M4 Sherman chassis.

Below: A rear view of the 160mm SP mortar carrier with kit laid out. The mortar is breech loaded by a crew of between four and seven men and achieves a high rate of fire with a maximum range of 9600m (10,500 yds).

Above centre: An Israeli 160mm Soltam M-66 mortar. In the firing position the wheels are angled inwards and the mortar can be turned through 360 degrees.
Above: An Israeli 120mm Soltam M-65 mortar in the travelling position. Although usually towed behind a truck, the M-65 can be broken down into four elements and loaded onto pack mules.

plastic ring is forced outwards and up against the groove, so that it makes a gas-tight seal. This device improves the efficiency and accuracy of the 81mm by a considerable amount, so that a maximum range of 5800m (6350yds) is quite possible.

Moving up from the 81mm level, several armies employ 120mm mortars, though in many cases they are considered as light artillery and are not manned by infantry. The average 120mm mortar can deliver a bomb weighing about 14kg (31lb) to ranges in excess of 6000m (6560yds) but the penalty for this is weight: many 120mm weapons have wheeled carriages incorporated in their baseplates and can weigh up to 600kg (1320 lb) or more.

One advantage of the 120mm calibre is that it allows the designer some space in which to make improvements to the basic bomb. While the standard 120mm bomb is much the same sort of projectile as the 81mm, the French Hotchkiss-Brandt company, long leaders in mortar development, have developed a rocket-boosted bomb. The bomb has a central rocket motor and a tail unit which is detachable. It is loaded and fired in the normal way, the explosion of the cartridge throwing the bomb high into the air. But the explosion also ignites a pyrotechnic delay element which, after some seconds of flight, blows off the tail unit and ignites the rocket motor. As the tail unit falls away it also unlocks a set of fins on the bomb which has, until then, been folded away. The bomb now accelerates under the impulse from the rocket motor, stabilised by the unfolded fins. With this PRPA RAP their standard 120mm mortar has its maximum range

Left: A FRELIMO unit in Mozambique. Light mortars are particularly effective weapons for guerrilla operations and ambushes since they can be broken down quickly and carried out of the combat area before enemy reinforcements can be called up.

Below left: The massive Soviet 240mm M-240 heavy mortar. The M-240 is breech loaded and fires 100kg (220lb) HE bombs to a range of 9700m (10,600 yds). Below right: A Soviet M-1943 120mm mortar. Six M-1943s equip the mortar battery of each motorised rifle battalion. Bottom: The seven-man crew of a Dutch YP-408 APC PW-MT variant deploys a French 120mm Brandt mortar.

increased from 4250m (4600yds) to 6550m (7100yds).

Another Hotchkiss-Brandt 120mm mortar design uses a rifled barrel, an idea which may seem to run counter to the basic idea of a simple weapon. The idea is not new; it was first put forward during World War I by an Australian officer, and a mortar to his design was developed by the US Army in 4.2-inch calibre and is still in use. In this type of mortar there is no need for fins on the bomb, since stability in flight is obtained by spinning the projectile just like a gun shell. The bomb looks rather like a shell, but with a short tubular extension at the base in which the ignition cartridge is carried and around which are the auxiliary cartridges. At the bottom of the bomb is a dished copper plate, smaller in diameter than the interior of the mortar barrel; below this is a flat steel plate. When the bomb is drop-loaded it slides down the barrel, leaving sufficient windage for the air to escape. When the cartridge explodes, the steel plate is driven forward against the dished copper plate and straightens it out so that it is now of slightly greater diameter than the bomb and digs into the barrel rifling. This causes the bomb to be spun as it passes up the bore, as well as making an efficient gas-tight seal, and as a result the 4.2-inch mortar was extremely accurate and had a range of just over 3660m (4000yds). Its present-day equivalent, the 107mm M30, uses the same system with an improved mounting and fires its bomb to 6800m (7400yds) range.

The Hotchkiss-Brandt rifled 120mm mortar is used by the French and several other armies. They approach the problem slightly differently, using a conventional copper driving band on the bomb which is pre-engraved to fit the rifling; this means that the loader has to fit the bomb into the rifling before he can drop it, which makes for a slight delay. In addition, there is also a rocket-boosted bomb for this mortar. With the standard spinning bomb the maximum range is 8135m (8850yds), while the rocket boost improves this to 13,000m (14,200yds).

Even larger mortars are employed by some countries. The Tampella company of Finland makes a 160mm weapon which fires a 40kg (88lb) bomb to a range of 10,000m (10,900yds), and a similar weapon, made by Soltam of Israel, who are linked with Tampella, is employed by the Israeli Army. The Soviet Army also uses 160mm mortars, though these are large and complex breech-loading weapons analogous to light field-artillery guns.

At the other end of the scale there is ample scope for

Above: The British 51mm is designed to be carried and used by one man and was introduced to replace the old 2in mortar. Early versions provided a monopod for support but this was deleted and a webbing gaiter fitted around the barrel for the firer to hold the barrel steady.

Modern Mortars: Weights and Ranges

Country of origin	Type	Total weight	Maximum range
Great Britain	51mm mortar	4.6kg (10.15lb)	800m (875yds)
Israel	60mm Tampella mortar	14.5kg (32lb)	2555m (2800yds)
Canada/ Great Britain	81mm light mortar	36.7kg (80lb)	5800m (6350yds)
Soviet Union	82mm M1937 mortar	56kg (125lb)	3040m (3325yds)
France	120mm RT-61 mortar	580kg (1280lb)	8135m (8900yds) PRPA RAP: 13,050m (14,750yds)
Israel	160mm Soltam M-66 mortar	1700kg (3750lb)	9600m (10,500yds)
Soviet Union	240mm M-240 mortar	3610kg (7940lb)	9700m (10,600yds)

Left: Two views of the French EMC 81mm mortar gun carrier. The EMC is a variant of the Panhard ERC armoured car fitted with an open-topped turret and armed with a Hotchkiss-Brandt 81mm breech-loaded mortar. The EMC is extremely mobile with a road speed of 110km/h (68mph) and a range of 950km (595 miles) and carries 72 HE bombs. The EMC's mortar can also be used in an anti-tank role, with APFSDS bombs.

Below: The French Panhard M3 APC can mount a wide variety of armament including the Hotchkiss-Brandt 60mm mortar. Like the EMC mortar gun carrier, the M3 provides a highly mobile platform and is fully amphibious.

very small mortars. The favoured calibre for this mortar type has been 60mm (51mm in the British Army) which is, today, almost the standard for what are usually called 'patrol' or 'commando' mortars. These can be carried by one man and put into action very quickly. The bomb they throw is only about 2kg (4.4lb) in weight and the maximum range is about 2000m (2190yds), but these weapons form a very useful support for the infantry platoon.

Another aspect of mortars which is currently witnessing much exploration is their mounting on armoured personnel carriers. The general policy is to strengthen the vehicle and mount the standard 81mm or 120mm weapon inside, firing through an open hatch. By using the standard weapon, and carrying the normal baseplate, it is possible to dismount it and use the mortar on the ground in the conventional manner should this be desirable. But some authorities are against this method, since they feel that the open roof of the vehicle makes the crew too vulnerable. They suggest that a better answer is the French 'gun-mortar', another Hotchkiss-Brandt invention. This is a 60mm or 81mm mortar barrel mounted in a recoil system and carried in the turret of an armoured car. It can be breech-loaded from inside the armour, so allowing maximum protection for the crew, and duplicates the performance of the standard 60mm or 81mm mortar. In addition, special ammunition can be provided to permit firing the weapon like a gun, so that anti-tank projectiles or mortar bombs can be fired on a flat trajectory. This gun-mortar has been adopted by several armies and it will probably find more adherents in the future.

RECOILLESS GUNS

The recoilless gun is really an attempt to retain the best features of artillery – shell weight, range – while losing the unattractive ones – weight and bulk – to produce a powerful weapon that can be easily transported. Though its first combat use (in Crete in 1941) was as a field artillery piece, it was later recognised that its special attributes made it the ideal anti-tank weapon, and it is in this role that it has prospered. Today the emphasis is on lightening it even more so that it can be operated by one man, and on doing away with the 'firing signature', the blast and flame from the rear of the weapon which gives away the gun's position when it fires.

When a conventional gun – whether a pistol or a howitzer – is fired, the weapon jumps backward with a force which is directly proportional to the energy expended in pushing the bullet or shell forward. This is caused by the explosion of the cartridge which exerts force in all directions: the sideways force is contained by the walls of the gun barrel, the forward force pushes the projectile, and the rearward force pushes on the weapon's breech and thus exerts the backward force – the recoil. Absorbing this recoil, in the case of the heavier types of weapon, is an expensive and complicated matter of applying hydraulic brakes to check the movement of the gun barrel and spring or gas recuperators to return it to its firing position in the mounting ready for the next shot. These recoil mechanisms are complex, expensive and heavy, and demand skilled attention for their maintenance and repair. If, therefore, it could be possible to make a gun which did not recoil then the cost and weight of this mechanism could be saved; moreover the gun mounting could be made much lighter since it would only have to support the weight of the barrel and not have to resist the recoil force.

Recoil systems first came into use in the late 19th century as guns began to become more powerful, and it was not until the early years of the 20th century that the advantages of a recoilless gun were first considered. The first innovator in this area of weapon design was an American, Commander Cleland Davis, who developed such a gun in 1914. Approaching the problem from basic principles he realised that a recoilless gun could be made by taking two identical guns and fixing them back to back. Firing them would thus mean that the recoil of each would cancel out the other's movement. Moving on from this he patented and developed a series of guns in which there was a central chamber and two barrels, pointing in opposite directions. The forward barrel was loaded with a conventional shell, the rearward barrel with a wad of grease and lead shot of the same weight as the shell. On firing the central cartridge, the shell was propelled forward to the target while the countershot went backwards and then disintegrated due to air resistance and centrifugal force.

The idea was taken up in Germany in the 1930s in order to provide the Luftwaffe with a heavy-calibre weapon with which to attack warships. The Rheinmetall company developed a 35cm gun carried by a Dornier Do-217 bomber. The gun fired a 635kg (1400lb) armour-piercing shell and the recoil was balanced by the rearward ejection of a 635kg (1400lb) steel cartridge case. The weapon was perfected by 1939 but the Luftwaffe had lost interest and it was never used. This was a minor setback, however, as the Rheinmetall engineers had taken matters a step further and had developed a recoilless gun which did away with the countershot.

The German engineers took the concept of a recoilless weapon to its logical conclusion. If the same weight of shot is fired at the same velocity in both directions the result is recoillessness; but the same result can be achieved by having the countershot half the weight of the service shot and ejecting it at twice

Previous page: A British paratrooper prepares to put a shell into the empty breech of a WOMBAT recoilless gun. Above: The first practical battlefield recoilless gun, the German 75mm gun used to considerable effect by paratroops on Crete in 1941. Below: The portability of the recoilless gun can be appreciated in this photograph of Viet Cong troops in South Vietnam at the end of the Tet offensive. Opposite page: A US 75mm gun in Korea (top) and 57mm gun in action at Okinawa (centre) and a British 95mm gun in 1945 (below).

the velocity, or a quarter the weight and four times the velocity. Carrying on in this way one can reach a point where the weight of the countershot is minute and the velocity extremely high, and the weapon is still recoilless. The German weapon utilised this principle in the extreme so that a column of gas ejected at high velocity acted as the countershot. The gun was built with a hole in the breech block and a venturi – a restricted throat to accelerate the flow of gases – in the hole. The cartridge case had its base pierced and plugged with a slab of plastic material. When the cartridge was fired this plastic remained whole for a fraction of a second, enough to allow the shell to begin moving up the barrel, after which it blew to pieces and released the high-pressure explosive gas through the rearward-facing venturi. In effect it was like having a rocket motor on the back of the gun pushing forward against the recoil.

Guns of this type in 75mm and 105mm calibre were secretly built and were unveiled by German paratroops in the battle for Crete in 1941. Their appearance led to experiments in Britain and the USA, and both countries had perfected recoilless guns before the war ended, though only the US Army managed to bring them into service. These differed from the German models principally in having cartridge cases with solid bases but which had their bodies perforated with holes. The propellant gas escaped through these holes into a chamber surrounding the gun, from which they were directed backwards through one or more venturis. Copies of the American 57mm and 75mm weapons are still in service with several countries, notably China and in South America.

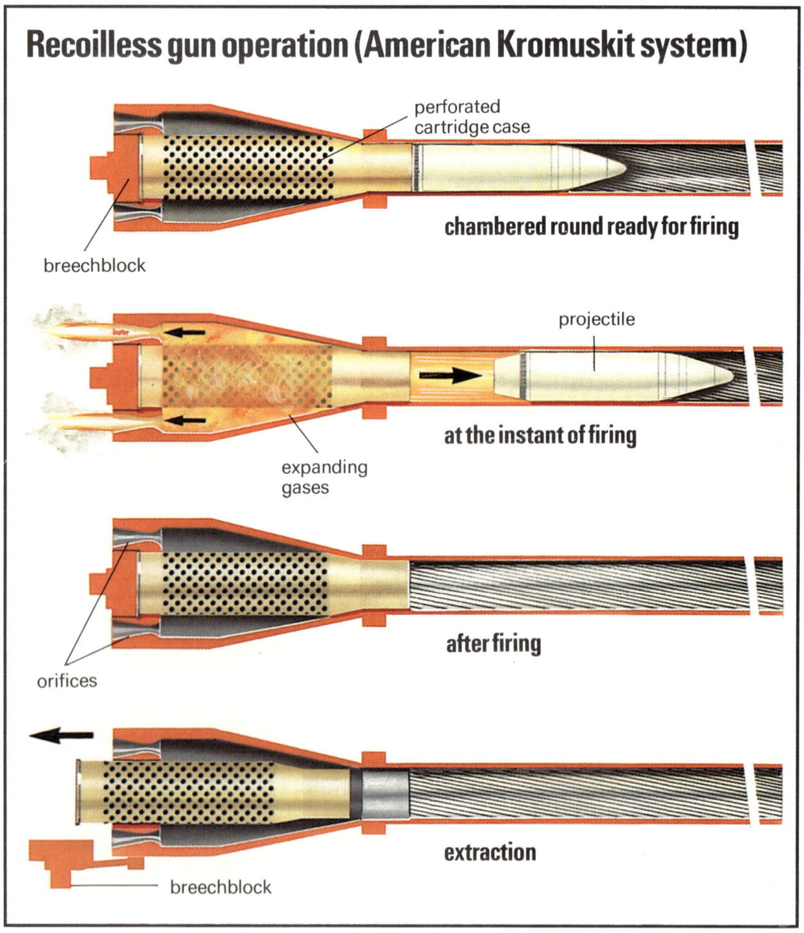

After World War II the British scrapped their experimental recoilless guns and began a fresh programme of development which led, eventually, to the 120mm BAT (battalion anti-tank) gun in the middle 1950s. This reverted to the German system of using a blow-out plastic base to the cartridge but improved it by building the vitally important venturi throat into the cartridge case itself. When the venturi throat was situated in the body of the gun, it accelerated the escaping gases, helping the recoilless effect, but the speed and heat tended to erode the throat so that after a few rounds had been fired the gun was no longer perfectly in balance and, indeed, would often begin to 'recoil' forwards. Placing the throat in the cartridge case meant that it was always of the correct size, since the case was only used once. The original BAT was a heavy weapon, weighing 1000kg (2200lb), but it was gradually refined, first into the MOBAT (765kg – 1680lb) which could be slung behind a jeep or Land Rover, and finally into the WOMBAT (295kg – 650lb). Besides a loss in weight, range was improved: while the BAT had an effective range of 500m (550yds), MOBAT's was 800m (880yds) and WOMBAT's 1200m (1300yds).

The Americans stuck to their perforated case system and issued first a 90mm and then a 105mm weapon. The 105mm gun turned out to be inaccurate and unreliable and was extensively redesigned; in order to distinguish the improved model the new gun was named the 106mm, and has since been taken into service by many armies throughout the world.

The principal role of the recoilless gun is that of an anti-tank weapon, though most have a secondary role as a general close-support weapon for the infantry. This came about because of the increasing size of conventional anti-tank guns designed to fire solid armour-piercing projectiles at high velocity. By 1945 these had gone beyond the size at which they were practical battlefield weapons; but at the same time improved methods of hollow-charge explosive attack on armour had appeared, methods which were independent of high velocity and therefore no longer demanded extremely powerful guns. What they did demand, though, was a large calibre so that the shell could carry a worthwhile charge of high explosive,

Above: Troops of the Gordon Highlanders receive instruction on a MOBAT recoilless gun. The MOBAT features a spotting rifle mounted above the main barrel. Above right: An FV 432 APC mounting a WOMBAT defends a river crossing during exercises in Germany.

and the recoilless gun was the obvious solution to this problem.

Another advantage of recoilless design was that it meant that if the weapon was light enough, there was no reason why it should not be fired from a man's shoulder, since there was no longer any recoil. The original German Panzerfaust of 1944 is often spoken of as a rocket, but it was in fact a small recoilless gun, and from this beginning has sprung a whole family of lightweight recoilless anti-tank weapons.

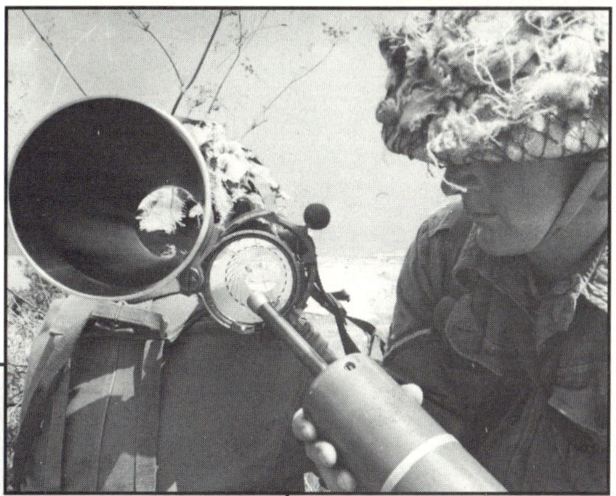

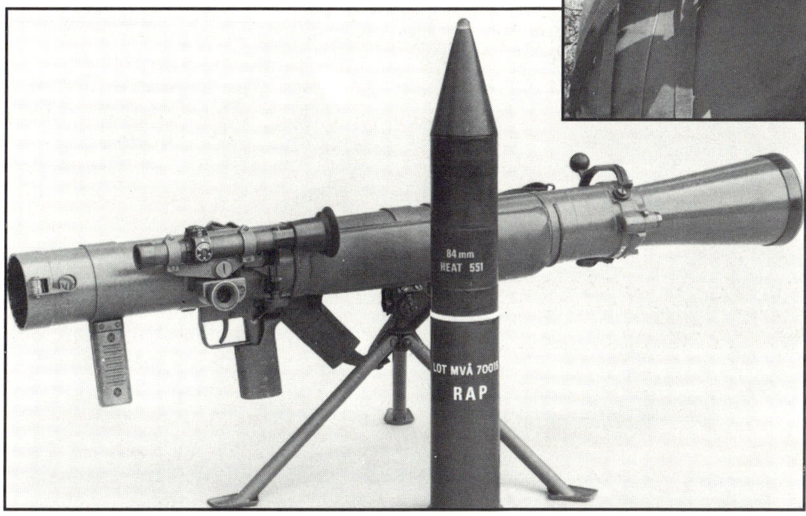

Above: A shoulder-mounted Carl Gustav is loaded with a HEAT charge. Left: A tripod-mounted Carl Gustav alongside a complete round. Below: A US-made 106mm recoilless gun is fired by a detachment of the Swiss Army. Given Switzerland's mountainous terrain ease of movement is an especially important tactical requirement.

Opposite page: Early British recoilless guns were fairly ungainly weapons, eventually discarded in favour of the BAT series. The complex venturi system of the 3.7in gun can be seen (above) while the man-portable 3.45in model is demonstrated (below). Below: Some of the force of the backblast of a WOMBAT recoilless gun is indicated on this Salisbury Plain firing exercise.

Above: A Swedish infantry man prepares to fire an M2 Carl Gustav. Above right: The Swedish Miniman short-range recoilless gun immediately after being fired.

Below: A soldier prepares to fire an Armbrust anti-tank weapon. The great advantage of the Armbrust over even lightweight recoilless guns is its 'enclosed' recoil which prevents the back blast from escaping into the open air where noise and flash can give away the firer's position.

One of the most effective is the Swedish 84mm Carl Gustav gun, used also by the armed forces of Nato and several other armies. This is exactly the same as the original German guns or the BAT, simply smaller and lighter so that it can be fired from the shoulder. The cartridge has a plastic blow-out base, venting through a venturi in the breech block, and the shell is a powerful shaped charge which can penetrate up to 400mm (16in) of armour plate at ranges up to 450m (490yds).

The recoilless principle has also been applied to disposable anti-tank weapons. The first disposable launchers were simply glass-fibre tubes with pre-packed rockets inside; the soldier placed it on his shoulder, fired the rocket, and then threw the tube away. But rockets, like recoilless guns, have one big tactical disadvantage: they emit a sheet of flame and a blast of hot gas and smoke from the rear when they fire, and this can reveal the firer's position for miles around. Moveover if the weapon is fired in a confined space or in front of an obstacle, the hot jet will wash back and injure the firer. This effect is known as the 'firing signature'.

Conventional recoilless guns such as the British WOMBAT and the US 106mm suffer from a high 'firing signature' and so have been replaced by more advanced weapons which minimise this weakness. Of this newer type of recoilless gun, a good example is the West German Armbrust, a simple tube carrying a central propelling charge located between two piston-heads. In front of one piston is the projectile, a shaped charge anti-tank bomb, behind the other piston is a countershot, a tight package of plastic flakes. When the cartridge is fired, the two piston heads are blown violently down the launch tube: the front piston ejects the bomb and sends it to the target, while the rear piston ejects the plastic flakes which disperse and flutter to the ground, doing no damage should they hit anything. The piston heads, though, are trapped as they reach the end of the tube and locked into place, so preventing the explosive gases or the noise of the explosion from reaching the open air. As a result there is no flame or smoke and the noise of the weapon firing is no more than that of a small-calibre pistol shot.

Once fired, the Armbrust is thrown away. It has a maximum effective range of around 300m (330yds), and the whole weapon weighs no more than 6kg (13lb). Since its appearance several other designs of countershot weapons have been developed, and these light disposable weapons represent the trend for the foreseeable future.

Armbrust recoilless launcher

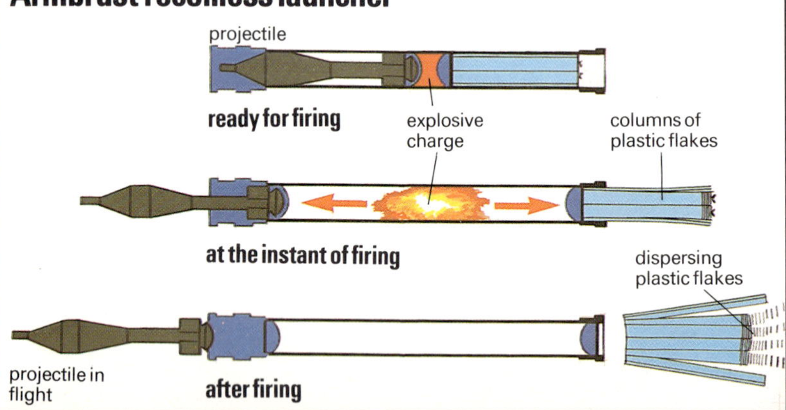

GRENADES

The Soviets call grenades pocket artillery' because of the additional firepower they give the infantryman. The hand grenade has seen much development since 1945, most of it aimed at improving fragmentation and controlling the lethal area so that the results can be more accurately predicted. The adoption of various types of pre-shaped fragments and the use of add-on fragmenting sleeves enables grenades to be 'tailored' to fit specific roles. The chemical grenade, containing either smoke or tear-gas, has also been greatly improved, largely due to the increasing use of these types in internal security situations.

Although the rifle has been (and remains) the infantryman's basic weapon, the soldier needs to supplement the rifle with additional firepower and the grenade is one of the simplest and most effective ways of doing this. Arguably, next to the rifle the grenade is the infantryman's most useful weapon. Grenades have a long history dating back to the 17th century when special units were formed as grenade throwers, being termed grenadiers. The modern grenade came into being during World War I when the need for a simple hand-thrown bomb achieved a new importance in short-range trench warfare.

While the Germans produced the famous Steilhandgranate or stick grenade in 1915, of more lasting importance was the British 'Mills Bomb' which as the No.36 became the prototype for the modern grenade. The No.36 consisted of an egg-shaped cast-iron body filled with high explosive, set off by a delayed firing system. This combination of case, filling and fuze are the three essential elements of the grenade, although they will vary according to the type and function of the grenade.

A number of different materials can be used for the case including cast iron, sheet metal, plastic or even cardboard. Grenades can be filled with high explosive – TNT or simple gunpowder – in an anti-personnel or anti-armour role or with chemical compositions for the purposes of smoke, illumination or signalling. There are two types of fuzing system. The first employs a pyrotechnic (or burning) delay element which after being activated by a striker will burn down to a detonator which in turn sets off the filler. The period of delay can vary between two and seven seconds. The second is the impact type (common in rifle grenades) in which the grenade explodes when it hits any hard surface. As a safety measure the impact grenade requires the force of being launched or thrown into the air to arm itself.

To operate the delayed-action grenade the thrower holds it in his throwing hand and removes the safety pin with his other hand. The grenade is then thrown in the desired direction; once released from the hand, the absence of the safety pin allows the safety lever to fly off which in turn allows a spring-mounted striker to shoot forward and hit the fuze initiator, thus setting off the short delay prior to detonation.

The best-known class of grenade is the fragmentation (or defensive) grenade. The British Mills Bomb was an early example, followed by the similarly-designed American Mk2 and the Soviet F-1. All three grenades feature cast-iron cases with distinctive deep serrations, both horizontally and vertically. The idea of the fragmentation grenade is for the high explosive charge to shatter the casing into a number of jagged fragments that fly out from the blast point in all

Previous page: Pinned down by enemy machine-gun fire, US Marines in Vietnam bombard a communist position with grenades before making their final assault. Above: An early version of the modern grenade in action with French troops in the trenches of the Somme. Right: The fragmentation provided by a British No.36 'Mills Bomb'. Below left: The Soviet RGD-33 stick grenade fitted with a serrated fragmentation sleeve. Below right: Priming a batch of No.36s.

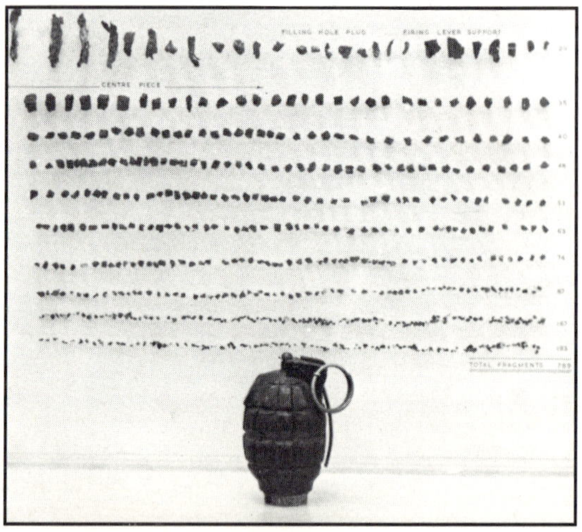

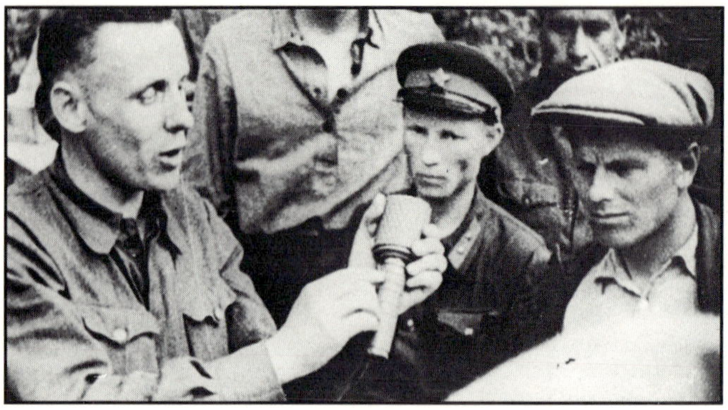

directions. While the range of such a grenade varies between 10 and 20m (11 and 22yds), isolated fragments can be thrown out to considerably greater distances so that the grenade thrower must be protected from the blast. Thus while the US Mk2 grenade can be hurled about 30m (33yds) at maximum by the average soldier, chunks of metal have been blasted up to 200m (220yds) on detonation.

The externally serrated case of the Mills Bomb type of grenade was subsequently found to be less than completely effective, as it did not fragment along the lines of the serrations but rather broke up at random. It has been improved upon by grenades which are serrated or notched on the interior, and fragment along the serration lines. By this method the fragmentation is far more uniformly even and the range and lethality of the blast can be controlled. Larger but fewer serrations produce fewer but larger fragments which carry further due to their greater momentum but at the same time have a less dense distribution pattern. Smaller serrations consequently produce smaller fragments which lack momentum and range but ensure that the target receives a dense distribution pattern. The manufacture of interior-serrated cast-iron bodies was found to be an expensive process and it was simplified by inserting a pre-notched wire coil behind a thin sheet-metal case. On explosion the metal case is vaporised allowing the wire fragments to fly outwards.

Besides the fragmentation grenade, the other important type is the blast or offensive grenade which is used in situations where a fragmentation grenade could endanger the thrower. During a rapid advance, for example, the soldier is not faced by the dangers of loose fragments flying through the air if a localised blast grenade is used. The blast grenade usually has a larger high-explosive charge than the fragmentation

Right: Ex-Sandinista commander Eden Pastora with a pair of US Mk2 'Pineapple' grenades. Grenades like the Mk2 do not break up evenly on detonation and have been replaced in US service with lighter, more efficient models such as the M26 (below right). The M26 utilises an internal pre-notched wire coil for fragmentation and has an improved throwing range and casualty radius.

Below: Grenades captured from Arab forces by the Israelis. The line to the left of the picture are Soviet F-1s, similar to the Mk2 and No.36, but fitted with the UZRG fuze protruding from the top.

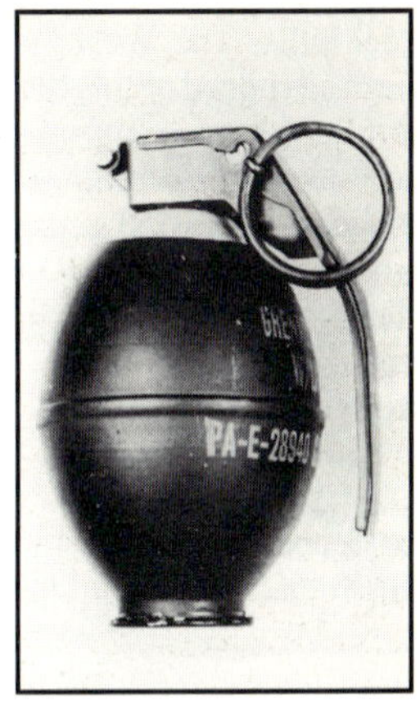

British No. 36M fragmentation grenade

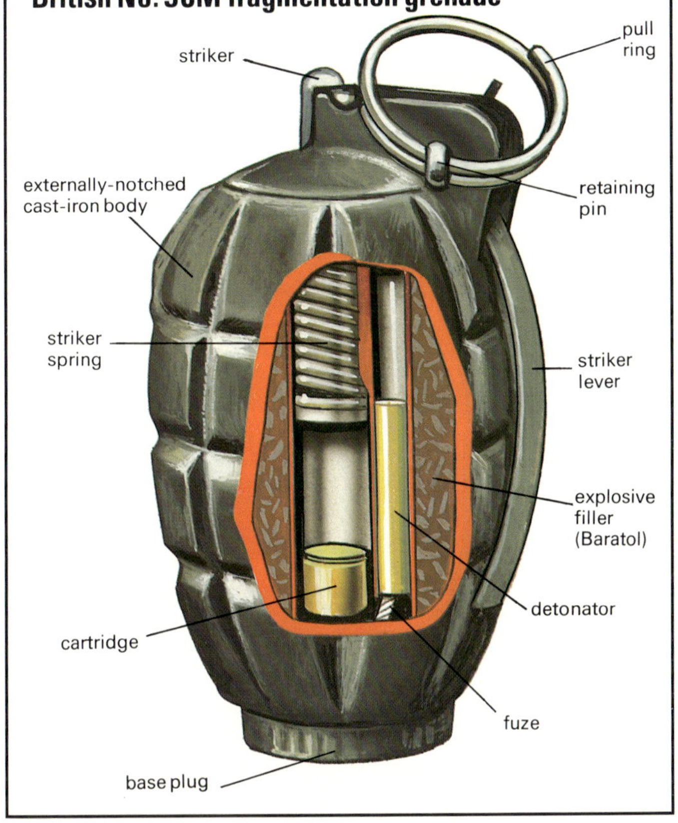

US M26 fragmentation grenade

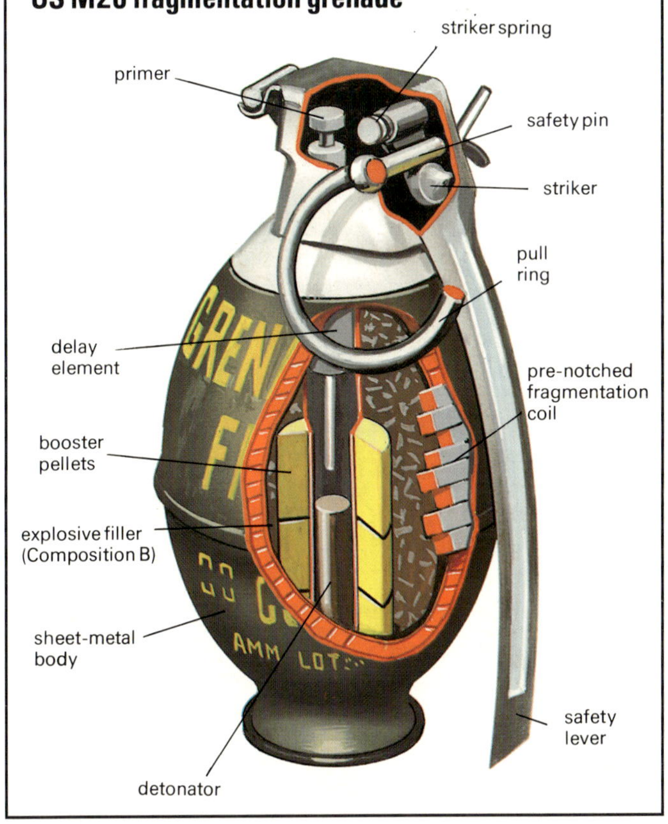

MG3

Calibre 7.62mm
Length 122cm (48in)
Weight 11.5kg (25.51lb)
Operation Recoil
Feed system Belt
Rate of fire 700-1300rpm
Muzzle velocity 756mps (2480fps)

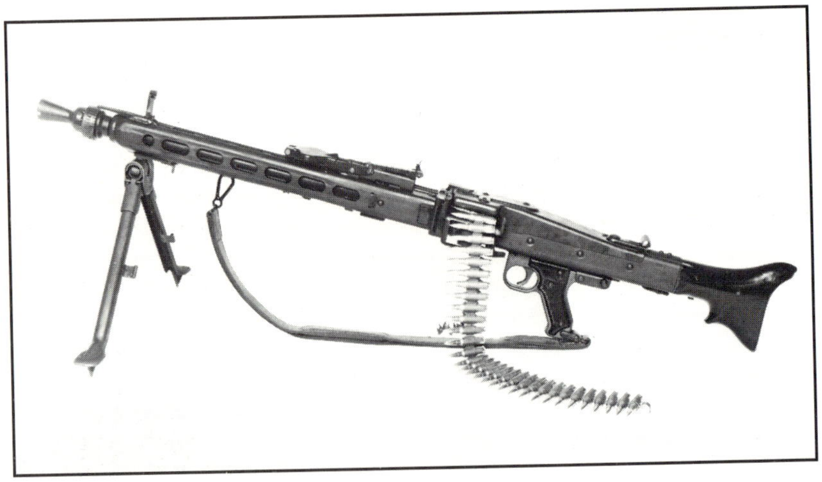

Above: The direct relationship of the MG3 to the wartime MG42 can be clearly discerned in this photograph.

perforated jacket which helps the airflow around the barrel and which is slotted on the right side to allow the barrel to be slipped out and changed. The MG3 can be fired from a bipod or from a tripod, and there are also a variety of specialised mountings for anti-aircraft defence or for firing from vehicles.

The Belgian FN MAG (Fabrique Nationale Mitrailleuse d'Appui Générale) was designed in the early 1950s as a gas-operated and belt-fed weapon, and has been widely adopted throughout the world; it is used in Britain as the GPMG L7A1 and the tank machine guns L8A1 and L37A2, and in the USA as the tank machine gun M240. It has also been made in other calibres, notably for Sweden as the 6.5mm M58 GPMG. A major factor in the FN company's success has been its excellent export record.

The mechanism of the FN MAG is rather unusual; it has a gas piston which is connected to the breech block by a toggle link. As the unit goes forward the block forces a cartridge from the belt into the chamber, and the toggle then drops across a bar in the gun body, so that any rearward movement of the bolt is locked by pressing the toggle against the solid bar. As the round is fired, so a small portion of the cartridge gas is tapped from a hole in the barrel and drives the piston rearward; this straightens out the toggle, unlocks it from the bar, and then withdraws the bolt to extract the empty case and begin the reloading cycle once more. A regulator in the gas cylinder allows additional pressure to overcome dirt or lack of lubrication in the mechanism and can also be used to adjust the rate of fire to between 600 and 1000 rounds per minute. As with other GPMGs the gun can be used on a bipod or on a tripod.

Left: The FN MAG can be carried ready-to-fire by the average infantryman and yet can still be utilised to lay down long-range interdiction fire. Below: A two-man crew of British infantry prepare to fire an FN MAG from a well-concealed position.

FN MAG

Calibre 7.62mm
Length 126cm (49.7in)
Weight 10.88kg (24lb)
Operation Gas
Feed system Belt
Rate of fire 600-1000rpm
Muzzle velocity 855mps (2800fps)

Right: The complete kit for the British FN MAG showing cleaning kit and spare barrels. When tripod mounted the stock is normally removed. Below: A member of the British Task Force blasts away with his FN MAG against Argentinian positions on the Falklands.

M60

Calibre 7.62mm
Length 111cm (43.7in)
Weight 10.43kg (23lb)
Operation Gas
Feed system Belt
Rate of fire 550rpm
Muzzle velocity 855mps (2800fps)

The American M60 was the result of development which began in 1944 by copying a German MG42; consequently it still uses some elements of the MG42 feed mechanism, though other aspects are somewhat different from the German gun. The bolt mechanism resembles that of the old-time Lewis gun, a gas piston carrying a vertical post which engages with the bolt and, by a curved slot in the bolt, rotates and withdraws the bolt on its backward stroke and then rotates it once more to lock into the barrel on the forward stroke. The bipod is, though, attached to the barrel, and the most important piece of equipment is the asbestos glove with which the gunner's mate removes the hot barrel, since there is no handle. Once again the gunner has to hold the gun off the ground until a new barrel and

Left: Prior to the real thing British troops tried out their FN MAG machine guns on Ascension Island to improve marksmanship and fire control.

bipod are fitted.

There is a tripod mounting for sustained fire, and there are also minor variant models of the gun for mounting in armoured vehicles or helicopters. Evaluation is currently in progress for a new lightweight model which has several improvements, notably the attachment of the bipod to the gun body and a handle on the barrel, making barrel-changing much easier.

Several armies are now beginning to question the whole GPMG concept, particularly since many have gone over to the 5.56mm cartridge for their standard rifle. It now makes sense to equip the infantry platoon with a light machine gun taking the same cartridge as the rifle, and several armies (notably the British and American) are in the process of adopting 5.56mm light machine guns. And as the 5.56mm bullet does not have the power for long-range machine-gun fire, some armies have stated their intention of staying with the heavier 7.62mm bullet, though perhaps adopting a lighter machine gun for the infantry squad. It now seems likely that the GPMG will be replaced by a dual system of 5.56 and 7.62mm machine guns.

Below right: A squad of M16-armed infantrymen support their M60 machine gun team during a battalion exercise. Below: The M60 in combat conditions – a sergeant in the 173rd Airborne Brigade defends his unit perimeter during fighting near Dak To in 1967. Like most other machine gunners in Vietnam he carries copious supplies of ammunition as well as a bottle of cleaning fluid held within his helmet band.

M60 GPMG

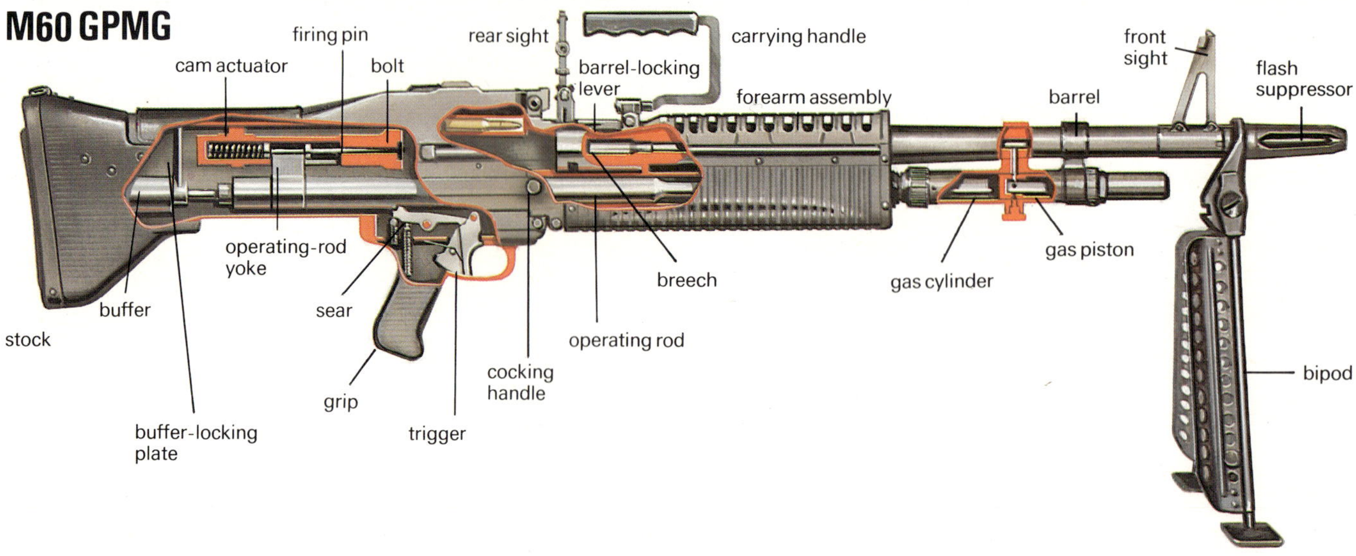

SOVIET MGs

The Warsaw Pact armies have probably had a greater diversity of machine guns in the past 40 years than any other armed forces, as various tactical theories have come and gone. They have always adhered to the policy of having a light machine gun for the rifle squad and heavier weapons for sustained supporting fire, although the PK model is similar to the Western GPMG. An interesting feature is the Soviets' gradual abandonment of old but reliable guns in favour of newer designs which are based on the highly successful and very widely used Kalashnikov AK-47 assault rifle: this simplifies many aspects of training and weapon maintenance.

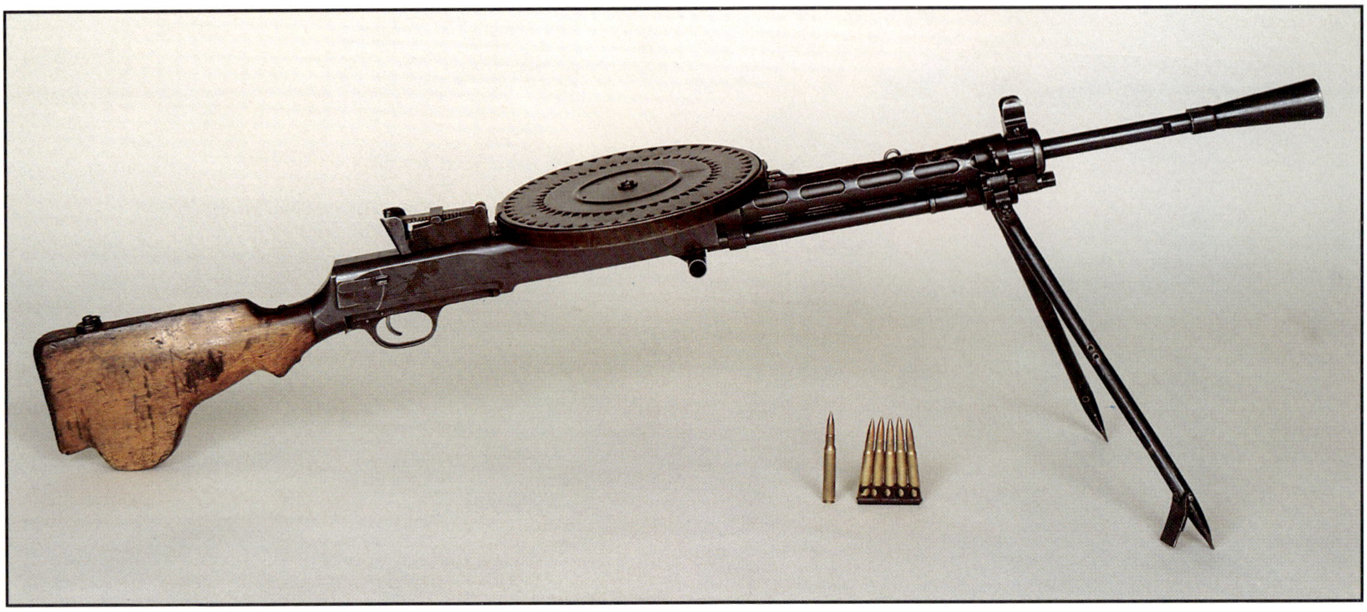

When the Soviet Great Patriotic War ended in 1945 the light machine guns of the Red Army were the 7.62mm Degtyarev DP and DPM. The DP dates from as early as 1926; it was a sound and robust weapon with several features that have been carried over to more modern Soviet machine guns. The DP was a very simple weapon: the locking mechanism employed only a few working parts and these, like the rest of the gun, were designed to be manufactured in as few operations as possible, using only simple machines and relatively unskilled labour. Once in the field the DP could be operated for long periods with only a minimum of maintenance or cleaning and it could take the hardest knocks that front-line service could inflict, which made it popular with front-line troops.

The DP, like any weapon, did have its faults. One was that it used the standard Soviet 7.62mm rimmed rifle cartridge. Ammunition had to be fed via a circular flat pan magazine, mounted over the receiver. This magazine, which gave the DP its distinctive appearance, was rather prone to distortion and was easily damaged in action. This fault only became obvious when the DP had been used in action over prolonged periods. Another major flaw lay in the fact that the barrel was fixed and therefore became overheated after prolonged periods of firing. Normally this would have been only a minor problem, but the operating spring was situated very close to the barrel and thus lost its tempering when the heat became excessive; the result was jams and feed problems. The Red Army had to put up with these problems until 1944 when the DPM, a modified version of the DP, was introduced.

Many of the best qualities of the DP were carried over to the DPM, including its robust and simple construction, but an important change was that the operating spring was moved to a new location and housed in a tube projecting from the back of the receiver. A distinguishing feature of the DPM was the introduction of a pistol grip for the trigger.

Neither the DP nor the DPM are retained in front-line use with any of the Warsaw Pact nations but they are still employed by various militias and reserve units. They are also frequently distributed to left-wing guerrillas who come into the Soviet sphere of influence. Few DPs are likely to be found today but the DPM is still around in large numbers.

After 1945 the DPM was used as the basis for another new Soviet machine gun. It evolved from the need for a type of light machine gun which could be used at company level to produce sustained fire over a much longer period than was possible with the DP and DPM. By careful re-working of the basic DPM it proved possible to employ a belt feed and further re-working enabled the barrel to be easily changed in action when it became hot. The ability to fit the 47-round pan magazine was retained but for this the entire belt feed mechanism had to be removed in one simple operation.

This new design was known as the RP-46 and for a while it was the standard company machine gun of the

Previous page: Left-wing troops man a position in the city of Beirut armed with an SG-43 machine gun and an AK assault rifle. Above: A DP machine gun, revealing its distinctive pan magazine and bipod. Below: The DPM is recognised by its pistol grip; this example is a Type 53, the Chinese version of the weapon. Bottom: A unit of Muslim troops parades through the streets of Beirut in 1980 armed with DP series light machine guns.

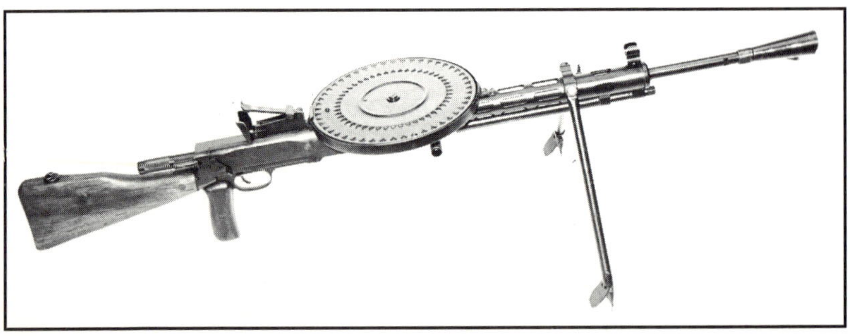

Soviet Army and many other Soviet-influenced armed forces, but its period in service was relatively short. For various reasons it was replaced by later designs but it is still widely used by the Chinese and North Korean Armies. The main reason for the passing of the RP-46 from Soviet use was that it continued to fire the old 7.62 x 54mm rifle cartridge. When the Soviet armed forces adopted the shorter and less powerful 7.62 x 39mm cartridge used by the AK-47 assault rifle, it made sense for a light machine gun to use the same ammunition as the standard service rifle and from this evolved the RPD light machine gun. In some ways the RPD was ahead of its day in design terms but in others it was something of a throwback. Among the better features was the special design of the disintegrating link belt feed: a belt feed can be a nuisance when the weapon is carried (it flaps about and tends to get caught up in clothing or undergrowth), but on the RPD the belt was carried inside a circular drum located under the receiver. This method is now widely employed by modern light machine guns such as the Belgian Minimi, but when the RPD appeared during the early 1950s it was a definite novelty. The RPD did not have a quick-change barrel, however, and to avoid serious barrel over-heating the gunner had to be trained not to exceed 100 rounds a minute. Other problems became evident only when the RPD was in service. For instance, the original model had a cocking handle that moved to and fro as the weapon was fired, and this was soon noted as a feature that had to be removed on later models. The handle snagged easily and, in addition, there was no cover over the handle's slot so dust and dirt soon found their way into the interior. In all there were five different production models of the RPD, not all of them associated with design problems, but it did occasionally make the interchanging of some spare parts between weapons a headache.

As with the DPM, the RPD is no longer a front-line weapon in the Warsaw Pact but it is widely used elsewhere. China and North Korea have both copied the RPD for their own use and many of these, along with ex-Soviet examples, have ended up in guerrilla hands all over the world.

The RPD could trace its ancestry back to the old DP light machine gun as far as the basic operating

Top: The RPD light machine gun with drum magazine. Above: A Syrian soldier stands guard on the outskirts of Beirut armed with an RPD. Left: Holding an RPD a masked Christian Phalangist takes up a rear-guard position on a patrol through the streets of Beirut. Both left and right-wing factions have made extensive use of Soviet weapons during the fighting in Lebanon. Below: A right-side view of the RPD.

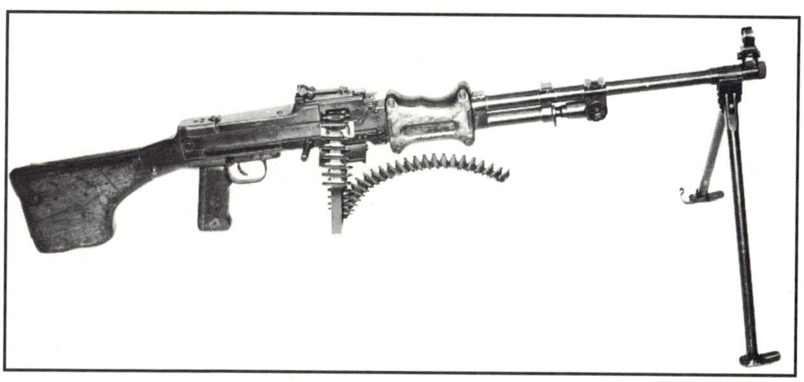

principles were concerned, but by the 1950s the Soviet Army began to favour the more modern and much simpler operating mechanism developed by Kalashnikov, the designer of the AK-47 assault rifle. The basic AK-47 design has proved to be one of the most widely-used and successful smallarms designs ever produced and it made a great deal of sense to adopt the same basic design attributes in a new light machine gun for use at section level. By the early 1960s such a weapon was already being issued to the front line units of the Red Army as the RPK.

The RPK may be regarded as a heavy version of the AK-47 assault rifle and it closely resembles the AK-47 in general appearance. Like the AK-47 it uses a distinctive curved-box magazine mounted under the receiver, but the magazine of the RPK is much larger; there are 30-, 40- and 75-round versions, the 75-round version having a drum magazine. The RPK has a light bipod fitted under the muzzle but otherwise the differences from the AK-47 are few. Thus the RPK has a fixed barrel and care has to be taken not to exceed a fire rate of 80 rounds a minute. In practice this is not a great drawback, since the RPK is used as a section weapon with the rest of the section using AK-47 assault rifles – so the combined firepower from a single section can be quite considerable without relying on the RPK alone. Further limitations are imposed by the use of the 30-round AK-47 magazine in the assault role, leaving the larger 40-round box or 75-round drum magazines for employment in the support role.

Many Soviet smallarms are somewhat rough in appearance due to the lack of care taken in the manufacturing stages – the Soviet philosophy is that if the weapon works it is good enough, and appearances will not help. But with the RPK, standards appear to have changed. The RPK is very well finished and such details as the chamber and bore are chromium-plated, a carry-over from the AK-47. This plating not only makes the weapon easier to clean but also tends to reduce wear and fouling during prolonged use. Another innovation on the RPK is that, whereas all previous Soviet light machine guns were designed, from the DP onwards, to deliver automatic fire only, the RPK has a fire selector mechanism, again a carry-over from the AK-47.

The RPK has an unusually-shaped wooden butt that is designed to allow the firer to hold the weapon well into the shoulder for stability when firing on automatic. This butt gives the RPK a distinctive appearance but there is a version with a folding butt known as the RPKS that is much used by airborne and other specialist formations. A new version of the RPK has recently emerged, the RPK-74, which fires the new Soviet 5.45 x 39mm 'miniature' cartridge as used on the AK-74 assault rifle. In general appearance it is identical to the RPK.

The RPK is now the standard Soviet light machine gun for use at section level but the Soviet armed forces

Above: The RPK is a heavy version of the AK-47 assault rifle, featuring an enlarged magazine (holding 40 rounds), folding bipod and a longer barrel. Below: The RPK in its primary role as an infantry support weapon at section level. Bottom: An Afghan guerrilla prepares to fire his RPK on a convoy of the regular Afghan Army held up in a mountain defile.

PK GPMG

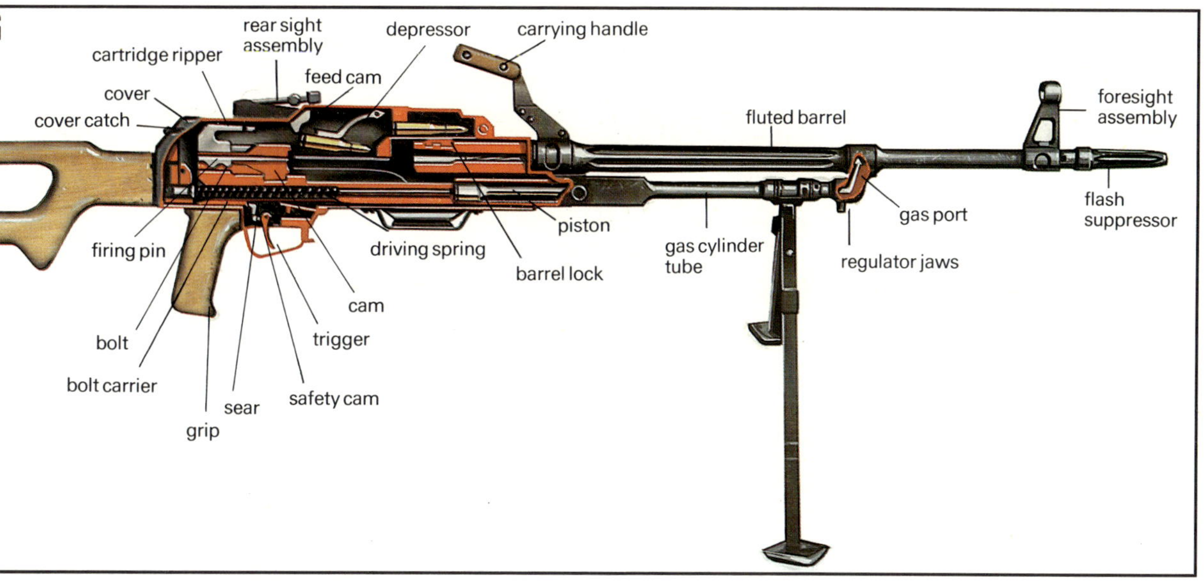

Left: An Afghan guerrilla leader stands by a PKM light machine gun fitted with a box for containing the ammunition belt. He is armed with a West German Heckler and Koch sub-machine gun.

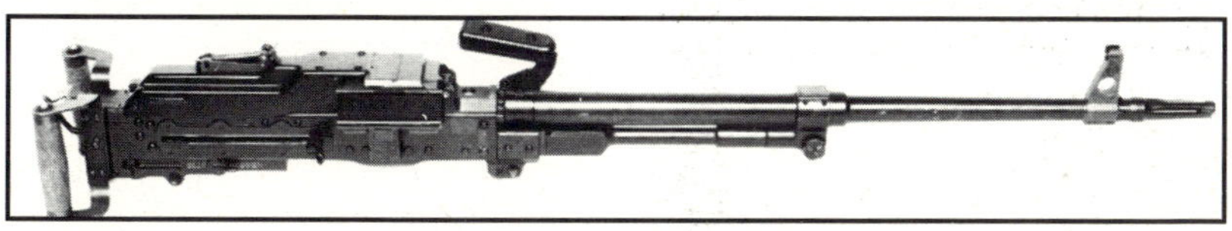

Below: An Afghan guerrilla fires a PKM from the shoulder alongside a downed Soviet helicopter. Inset: A side view of the PKM which shows its distinctive carrying handle and cutaway frame stock. Bottom: The PKB variant is distinguished by twin spade grips and a butterfly trigger.

also use a heavier weapon as a general purpose machine gun at company level and above. This is the 7.62mm PK, a weapon that can be fired from a bipod for the assault role or from a heavy tripod for use in the support role. It fires the old 7.62 x 54mm rimmed rifle cartridge that provides the weapon with a longer potential range than the lighter weapons, but the use of this elderly cartridge must have presented the designers with more difficulties than the use of a rimless one would have entailed. As it turned out the designers did a very good job, for the PK and its derivatives must be rated as being among the best machine gun designs in service anywhere in the world.

The basic PK uses the rotating bolt of the AK-47 and RPK combined with some of the best features of other machine gun designs. It first appeared in 1964 and since then its use has spread to all arms of the Soviet and Warsaw Pact forces. For a machine gun the PK is remarkably light and the weapon has only a few moving parts that are easily removed and replaced for repairs or cleaning. It has a changeable barrel for prolonged firing and ammunition is fed using 100-, 200- or 250-round belts. The ammunition belts are usually fed into the gun from a box carried under the receiver (as on the RPD), but this box is not always needed once the gun is mounted on a tripod or a fixed mount.

The PK is just one of a family of weapons and in its basic form it has a fluted barrel and weighs about 9kg (20lb). When mounted on a tripod for prolonged fire it becomes the PKS, while the PKT is a special version for co-axial mounting on armoured vehicles – this version has a special firing solenoid and a heavier barrel. The PKM is a generally simpler and lighter version of the basic PK and when this model is mounted on a tripod it becomes the PKMS. To complicate matters further, there is a version known as the PKB or PKMB which is a PKM mounted on a tripod with the normal trigger arrangement replaced by twin spade grips and a butterfly trigger.

The PK family of weapons are very simple and effective weapons. The rate of fire is not excessively high, being of the order of 650 to 720 rounds a minute, and so ammunition expenditure need not be over-lavish even during prolonged periods of firing. Recoil forces are relatively low so the weapon can be easily

Left: The SGM, mounted on a heavy two-wheeled carriage fitted with a detachable shield. Despite its cumbersome mounting the SG was a highly effective machine gun capable of accurate sustained fire at extended ranges. The PK series of general purpose machine guns has replaced the SG in front-line Soviet Army service.

handled, even when fired from the bipod. Barrel changing is quick and simple.

The PK family is the result of gradual development from a variety of sources to produce a weapon for a specific purpose. As a general purpose machine gun it is now one of the finest of its type in use anywhere, but there are definite limitations on the performance of a machine gun firing a standard rifle cartridge in the field; for some types of target and application heavier weapons are required. The Soviets have produced many heavy machine guns over the years but the only one that is still likely to be encountered is the Goryunov M-1943, the SG-43. This machine gun was originally designed to be the heavy counterpart to the DP and DPM light machine guns when it was first introduced in 1943, but these days it may be classified as a really heavy machine gun. The SG-43 fires the standard 7.62 x 54mm rifle cartridge and it was designed by Goryunov to replace the venerable M-1910 Maxim machine gun.

The SG-43 carried over several features from the old Maxim, including a wheeled carriage that used a shield to protect the firer. This made the SG-43 weigh in at about 40kg (88lb) which made a hefty load to handle, but the gun itself weighed only 13.6kg (30lb). The SG-43 was belt-fed and air-cooled and had a quick-change barrel device for use during prolonged periods of fire. It was certainly robust and reliable but the overall weight factor led to several changes being introduced. One was to the gun itself which became the SGM, recognisable by the introduction of a fluted barrel. The wheeled carriage was replaced by a lighter tripod which reduced the weight to 27.4kg (60.3lb).

The SG-43 and the SGM are no longer used as front-line weapons by the Warsaw Pact nations but many are still in service with countries around the world, ranging from Egypt to China. The SG-43 or the SGM are virtually standard hand-outs for any nation that comes into the Soviet sphere of influence and although not ideal guerrilla weapons, they are often encountered in use by such formations. Perhaps the greatest use of the SG-43 and SGM is made when the weapons are placed on anti-aircraft mountings, for these machine guns can fire very effective armour-piercing rounds (as can the PK family). Special tank versions known as the SGMT or SGMB are still widely used on some of the older Soviet tank models.

Below: An Afghan guerrilla machine-gun crew awaits the order to fire its SG-43. Lacking the refinements of the SGM, the SG-43 is nonetheless an ideal weapon for the long-range fire-fights encountered in the Afghan conflict.

Soviet machine guns

Model	Calibre/ cartridge	Length	Weight (empty)	Ammunition feed	Rate of fire	Muzzle velocity
DPM	7.62×54mm	127cm (50in)	9.1kg (20.1lb)	47-round pan	600rpm	840mps (2756fps)
RP-46	7.62×54mm	128cm (50.4in)	13kg (28.7lb)	250-round belt	600rpm	840mps (2756fps)
RPD	7.62×39mm	104cm (40.9in)	7.1kg (15.7lb)	100-round belt	700rpm	700mps (2297fps)
RPK	7.62×39mm	104cm (40.9in)	5kg (11lb)	30/40-round box, 75-round drum	660rpm	732mps (2402fps)
PK	7.62×54mm	116cm (45.7in)	9kg (19.8lb)	100/200/250-round belt	650-720rpm	825mps (2707fps)
SG-43	7.62×54mm	112cm (44.1in)	13.6kg (30lb)	250-round belt	650rpm	800mps (2627fps)

STEN & STERLING SMGs

It is a strange fact of history that Britain, which steadfastly refused to contemplate the use of 'gangster guns' before 1939, eventually became a major supplier of sub-machine guns to the rest of the world. Over three million Sten guns were made during the war and distributed freely throughout the occupied countries of Europe, from whence they found their way all around the world.

In post-war years the Sterling, which replaced the Sten in British service, has been adopted by over 90 military forces and has been manufactured under licence in several countries. The Sten introduced the concept of cheap, almost 'disposable' weapons, while the Sterling showed how a simple design could be made robust and reliable.

Previous page: A British soldier prepares to cock his Sterling sub-machine gun during an incident in the fighting for Aden in the 1960s. Above: An early experiment in fitting infra-red equipment to smallarms, here a Sten with silencer.

In 1940 the British Army's requirement for a new, cheap, simple-to-operate and reliable sub-machine gun was met by a weapon that by the end of World War II had seen action on practically every battlefront in every theatre of the war. The Sten sub-machine gun set new standards of cheapness and mass production for smallarms and at one stage of the war Stens were being turned out for the equivalent of £3.50 apiece, with one special factory alone producing over 20,000 models a week in 1942.

The first Sten to appear in June 1941 was the Mark I which, despite attempts to keep its design as unrefined as possible, was soon replaced by an even simpler version, the Mark II. The only machined parts on the second model were the barrel and the bolt, while the body was a simple steel tube, the magazine housing a pressing, the pistol grip a steel stamping and the stock a simple piece of tubular steel. The design of its successor, the Mark III, reduced the manufacturing process even further, providing for a fixed barrel and one-piece body and barrel casing. A further variant on the Mark IV version, intended for parachute use with a folding stock and cut-down body, was tried but due to operational problems was never introduced into service. A 1943 silenced version of the Mark II proved highly successful, however, and equipped Resistance fighters all over Europe.

The final version of the Sten, the Mark V, was introduced in 1944; it was better finished, had a wooden stock and handguard and mounted a bayonet, although no attempt had been made to rectify the Sten's main weakness, its habit of jamming at critical moments. Jamming could result from only slight damage to the long magazine which affected the feed. In spite of this weakness, provided the weapon was looked after and the magazine in particular treated with care, the Sten worked and kept on working when many more expensive and elaborate weapons failed due to the effects of dirt, cold and mud.

The Sten provides a cyclical rate of fire of 540 rounds of 9mm ammunition per minute and operates on a simple blowback method. The loaded magazine is inserted into the magazine housing assembly and the weapon is prepared for firing by pulling back the cocking handle, thus compressing the return spring. When the trigger is pressed the breech block is released and shoots forward due to the pressure from the spring, driving a cartridge into the chamber (by means of feed ribs on the block); the cartridge is then detonated by the firing pin. Inertia and pressure from the spring keep the breech block closed until the bullet has left the barrel when the remaining pressure drives the spent cartridge case and breech block assembly back. The cartridge case is ejected, a new round pushed in line for feeding, and the firing process repeated.

An estimated three-and-a-half million Stens have been produced and the weapon has seen action all over the world since 1945 with a number of armies. Apart from its use with the British Army in Palestine in 1948 and in Malaya in the 1950s, it has been copied in Germany, China, Belgium, Argentina and Indonesia and was also used by the Viet Minh against the French in Indochina (1946-54).

Despite the Sten's success and massive production levels, not everyone had been happy with it and several attempts were made to put forward an improved sub-machine gun. One of the first to be

The Sten Mk I was characterised by a rudimentary forward grip and a cone-shaped flash hider.

A modified version of the Mk II was the Mk IIS, fitted with a silencer. It was particularly suitable for undercover work.

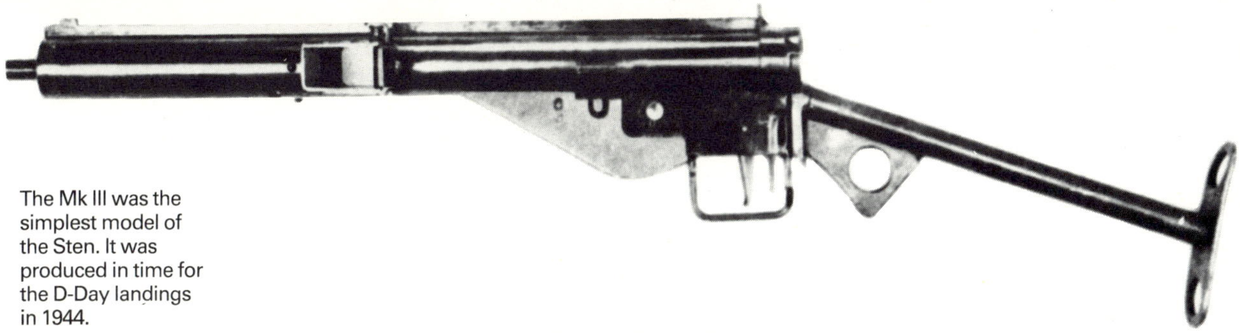

The Mk III was the simplest model of the Sten. It was produced in time for the D-Day landings in 1944.

The Mk VI was the silenced version of the Mk V, perhaps the best of the many Sten models to see service.

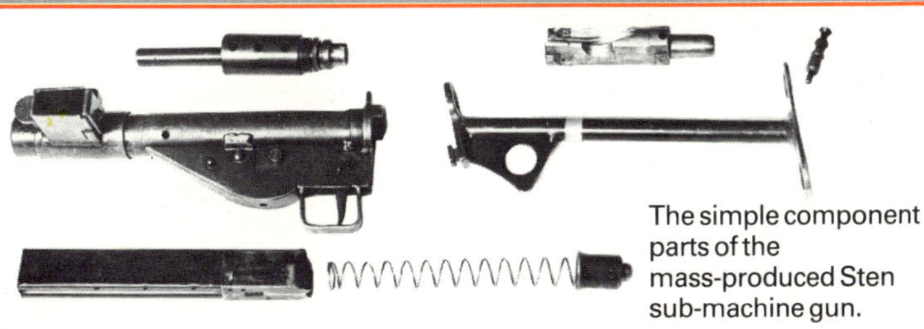

The simple component parts of the mass-produced Sten sub-machine gun.

Sten Gun Mk II

Calibre 9mm
Length (overall) 76.2cm (30in)
Weight (loaded) 3kg (6.62lb)
Rate of fire Cyclical 540rpm
Maximum effective range 90m (100 yds)
Magazine 32-round box
Muzzle velocity 390mps (1280fps)

considered was the Veseley, designed by a Czech engineer who had come to Britain in 1939 to help with production of the Besa machine gun. Unable to return to Czechoslovakia, he stayed in Britain and in 1940 submitted designs for a new sub-machine gun. These were rejected since the Sten had just been approved, but he persevered, building a gun at his own expense which was tested in November 1942. It was a simple blowback weapon, but well finished, and the magazine was of particular interest; it featured two 30-round columns, one in front of the other. Feed was initially from the front column, and when this was exhausted it fed from the rear half. The Royal Navy showed interest and Veseley made various minor changes, continuing tests until 1945, but although the gun was not unsatisfactory there were, by then, other promising designs and the Veseley was quietly abandoned.

Leaving aside several projects which were obviously hopeless, the next one to show promise appeared in September 1942 when George Patchett demonstrated a new design to a military panel. Like the Veseley, it was a simple blowback gun, but since Patchett intended it solely for firing from the hip it had no butt or sights. Trigger and pistol grip were situated at the point of balance and it featured an ingenious trigger mechanism with a simple thumb-operated change lever to provide single shot or automatic fire. The panel recommended that Patchett fit it with a butt and sights, whereupon it would be tested against the Sten and the Welgun, a sub-machine gun developed by the Special Operations Executive (SOE). The trial duly took place in February 1943 and the Patchett was reported to be inaccurate over 150 yards and prone to stoppages from mud and dirt.

After further modification another test was run in September 1943, this time with more contestants including the Australian Owen gun, a modified design based on the Sten, an Australian Austen, and the Welgun. This time the Patchett came second to the Owen, which, considering the Owen was probably one of the most reliable sub-machine guns of World War II, was very satisfactory. Despite Patchett's success it was announced that production limitations made it unlikely that any new design of sub-machine gun would be needed, but if one were, then the Patchett looked closest to meeting the requirement.

In January 1944, however, the War Office issued a set of specifications for future weapons, one of which was for a sub-machine gun; in brief, it had to weigh 2.7kg (6lb), be in 9mm calibre, be capable of putting

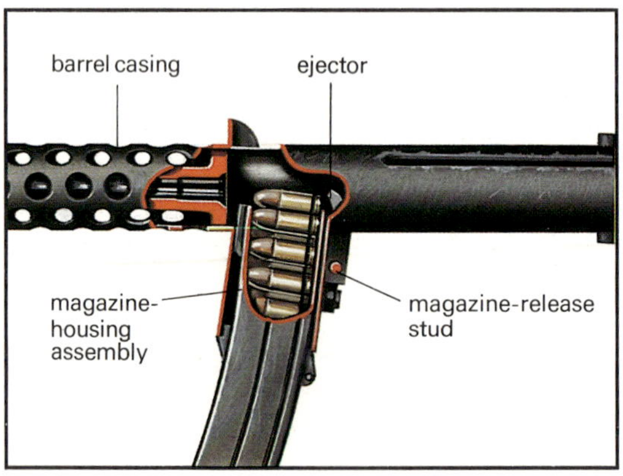

Above: A radio operator advances armed with a Sterling SMG. The 9mm Sterling replaced the Sten as the British Army's sub-machine gun in the 1950s. Although not a radical advance on the Sten the Sterling has proved highly effective as its successful export record has shown.

Sterling L2A3 SMG

Top: A member of the Royal Ulster Constabulary surveys the damage of an IRA bomb-blast, armed with the Police Carbine version of the Sterling. Above: A sergeant of the Royal Marine Commandos carries a Sterling while on patrol during the fighting in Brunei in 1962.

five single shots into a 30cm (one-foot) square target at 90m (100 yards) range, and have a rate of fire not exceeding 500 rounds per minute. The Patchett seemed a likely candidate to meet this requirement and a new model, with helical ribs on the bolt to scour out dirt and dust, was tested. As a result, on 12 January 1944 20 Patchett guns were ordered from the Sterling Armament Company, followed by another order for 100 for 'special troop trials'. Full trials were carried out in September 1944 and the Ordnance Board announced that 'with regard to functioning, accuracy, endurance and penetration, it is considered that the Patchett machine carbine is suitable for service.' The 100 for 'special troop trials' were actually issued to the Parachute Regiment and saw action in the battle for Arnhem in September 1944.

After 1945 the pressure for a new weapon waned and more applicants came forward with designs. Among the best of these was a design from BSA Ltd, which featured an unusual method of cocking by pushing the foregrip forward and a magazine which could be hinged forward for stowage or to clear stoppages. Unfortunately it proved expensive, which reduced the number the War Office was prepared to buy for trials, and when a demand for a bayonet was added to the specification, the peculiar cocking arrangement made this addition difficult.

In 1951 further trials were held with the BSA, a new Patchett (with some slight changes to the firing mechanism), and a design from the Royal Small Arms Factory at Enfield. The Patchett demonstrated its superiority over the others in every respect; it provided the desired slow rate of fire, could be stripped without tools, and continued to work in spite of dust and mud. The Ordnance Board recommended that the Patchett should be adopted if the forthcoming EM2 7mm rifle proved unsuitable in the sub-machine gun role. In the event the EM2 rifle was not adopted. The new FN FAL rifle could not function in the sub-machine gun role, and the Patchett became the official British sub-machine gun on 18 September 1953.

Although referred to as the Patchett in the Army Estimates for 1954-55, it is popularly known as the Sterling, since it has always been manufactured by the Sterling Armament Company of Dagenham. In addition, however, over 160,000 were manufactured by the Royal Ordnance Factory Fazackerly; indeed, the Ministry of Defence used Patchett's patents in these guns and, after much argument, Patchett was forced to issue a writ against the Crown to obtain recompense. After a long drawn out struggle he won his case and was awarded £116,975 in June 1966.

The standard production model L2A1 as used by the British and other armies is a blowback weapon feeding from a side-mounted curved box magazine. The Sten's principal disability lay in its magazine, and the Sterling magazine has been carefully designed to avoid stoppages. The cartridge follower carries roller bearings to reduce friction, and the entry is angled so that the cartridge cannot align with the firing pin until it has entered the gun chamber. A folding metal butt is provided, and a pistol grip at the point of balance. The barrel is surrounded by a perforated jacket which acts as a forward grip.

A later version of the Sterling, the L2A2, appeared with several modifications to the original design. The L2A1 featured a number of parts which could be used as stripping tools while the L2A2 dispensed with this facility but included a forward finger guard, a modified chamber and rear sight, and a strengthened butt. The current standard service model, the Sterling Mark 4 (L2A3) includes further modifications including a redesigned wooden butt and a chamber modified to the Nato standard. Further variants include the special Police Carbine and the L34A1 silenced model.

The latest Sterling design is the Mark 7 Para Pistol, a specially shortened version for use in confined spaces. It can be supplied either firing semi-automatic or fully automatic, and there is the option of firing from an open or closed bolt, the latter giving improved accuracy.

The current version of the Sterling has proved a reliable, well-made and effective weapon in a wide range of conditions and has been sold to over 70 armed forces and defence units throughout the world. Its accuracy has proved far superior to that of the Sten and although a more expensive weapon, it retains the Sten's simplicity of operation and ruggedness while achieving a far higher level of reliability – a tribute to the soundness of its basic design.

The first of the Sterling models was the 'Patchett' SMG, produced in 1944 and used at Arnhem in small numbers.

The Sterling without its folding stock and with a 10-round magazine instead of the standard box.

The Sterling is sometimes fitted with a wooden rather than the more usual collapsible metal stock.

The silenced version of the Sterling, the equivalent of the silenced Sten model of World War II fame.

The Mk 7 Para Pistol, a cut-down version of the standard Sterling, used by airborne troops and commandos.

Sterling L2A3 SMG

Calibre 9mm
Length (stock extended) 69cm (27in)
Weight (loaded) 3.5kg (7.7lb)
Rate of fire Cyclical 550rpm; practical 102 rpm
Maximum effective range 200m (220 yds)
Magazine 34-round box
Muzzle velocity 390mps (1280fps)

MODERN SMGs

The Sten SMG (sub-machine gun) and similar wartime weapons were effective but scarcely elegant, and the post-war years saw innumerable attempts to develop replacements for them. But the vast number of war-surplus SMGs meant that there was little demand for new models until the Israeli Uzi showed how compactness and reliability could be built into the same weapon package. Several designs have appeared in the 1970s and 1980s, but at the same time the military use of the sub-machine gun has declined in favour of the assault rifle. It looks as if in the future the anti-terrorist role is the principal task for sub-machine guns.

Previous page: A streetfighter in El Salvador with an Uzi sub-machine gun. Right: An IRA checkpoint in Londonderry in 1972. The man on the left is armed with a Thompson M1928A1, a predecessor of the more utilitarian M1A1. Below: A member of the Swedish Army on peacekeeping duties in the Sinai in 1976. He is armed with a Carl Gustav Model 45, one of the more successful postwar sub-machine gun designs.

World War II brought the sub-machine gun into prominence – by 1945 several million had been made. Designs such as the Sten, the Soviet PPS-43 and PPSh-41, the American M3A1 and the Thompson M1A1 (originally developed as a civilian weapon in the 1920s) lacked refinement, but were rugged weapons and had a long postwar career.

The first new designs to appear after World War II were from Scandinavia. Towards the end of the war the Swedes had decided to produce a sub-machine gun of their own design and manufacture and the Carl Gustav State Arsenal developed the Model 45 which has remained in use ever since. It has sold widely to other countries, has been manufactured in Egypt as the 'Port Said', and licence-built in America by Smith & Wesson for the US Navy. The mechanism was simple, a blowback system largely derived from the Sten gun while the body and barrel jacket were of stamped steel, riveted together. The Model 45 featured a simple tubular butt, hinged to fold forward alongside the body and its 36-round magazine was designed and manufactured so as to avoid the feed jams which plagued most wartime sub-machine guns.

The Madsen Company of Denmark produced an unorthodox sub-machine gun – also known as the Model 45 – which in most respects was an enlarged

M3A1 Sub-machine gun

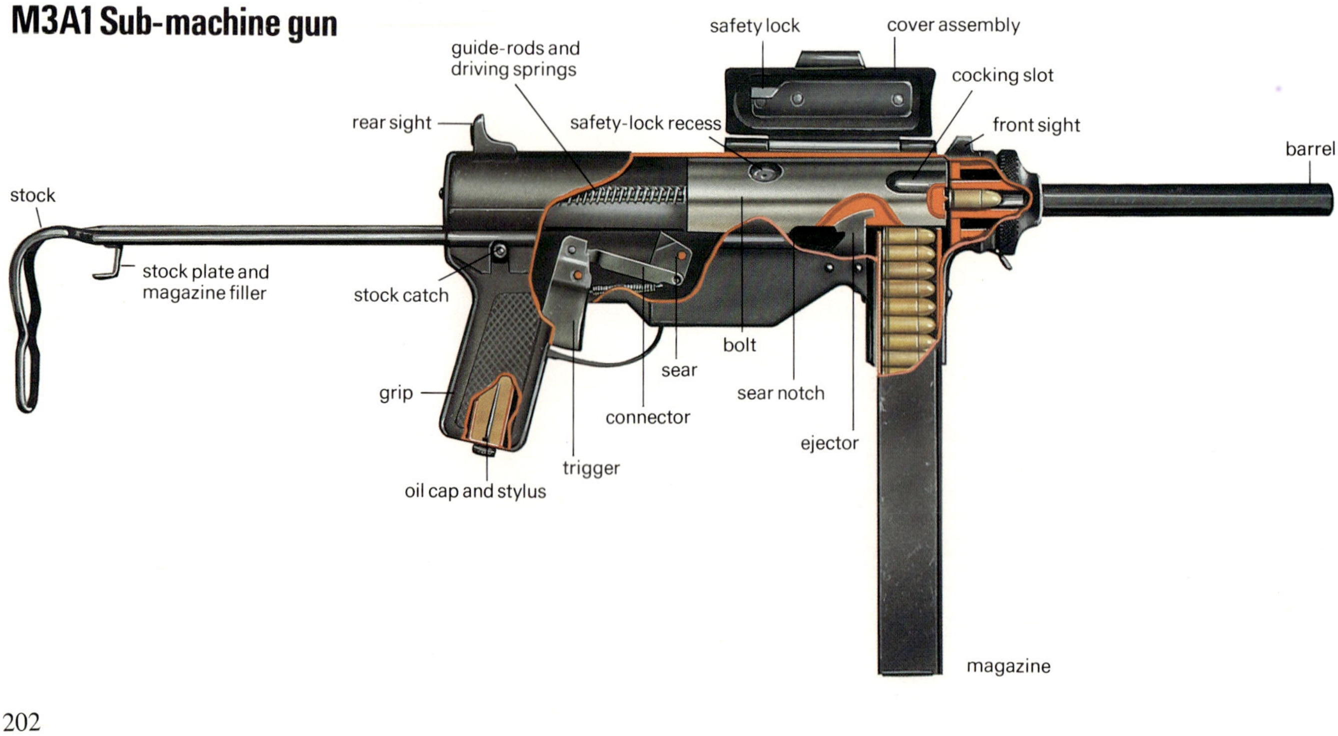

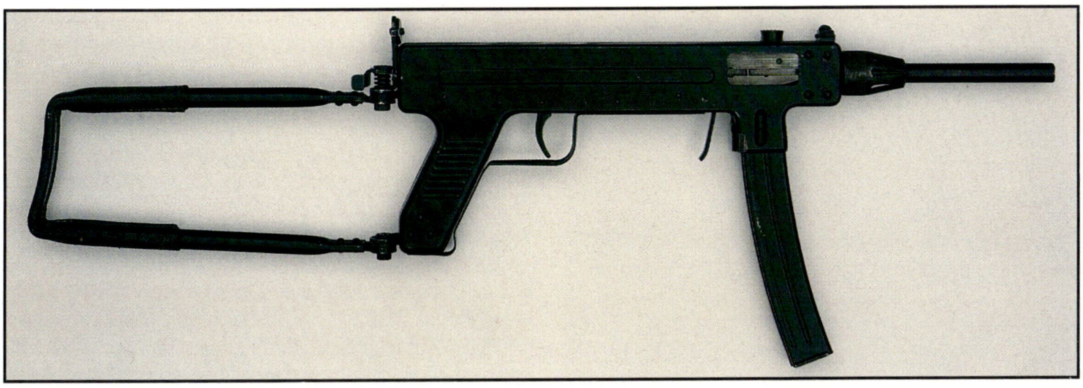

Below: A Chad government soldier armed with a MAT-49 sub-machine gun. This compact and robust weapon was produced in France, and used 9mm Parabellum ammunition. It featured a magazine assembly that could be folded forwards underneath the barrel when not in use. The MAT-49 equipped the French Army and was used extensively in Algeria.

automatic pistol. The Model 45 was not popular, however, and it was soon replaced by the Model 46, a conventional blowback type which was an instant success, exported throughout the world, particularly to South America. A unique feature of the Model 46 was that the body was made in two halves, hinged at the rear and locked together by the barrel-retaining nut at the front. By unscrewing this nut the two halves could be hinged open like a book and the working parts exposed for cleaning and repair.

An interesting feature of the Madsen was the safety grip, positioned just behind the magazine housing, which with the magazine acted as the forward hand guard and ensured that the gun had to be fired two-handed. The Madsen was subsequently updated as the Model 50, and during firing demonstrations in 1950 it impressed the British delegation present sufficiently to recommend that it replace the Sten, although in the event, the British Sterling was the replacement.

The French produced a 7.65mm calibre sub-machine gun before World War II – the MAS-38 – but this weapon did not have a long service life as it was replaced by the MAT-49, although a number of examples were used by colonial forces in Indochina. The MAT-49, made by the National factory at Tulle,

Above: The Danish Madsen Model 50. This sub-machine gun works on the standard blowback principle and can fire single shots as well as full automatic. The Madsen has a similar folding stock to the Swedish Carl Gustav, although a rather unusual feature is the safety grip mounted just behind the magazine housing.

Below: The Owen was produced during World War II as a useful jungle warfare weapon for the Australian Army. The camouflage pattern shown here was typical for this type, most being camouflaged in some way or another. A highly reliable sub-machine gun, the Owen was widely used in the anti-insurgent campaigns after 1945.

employed 9mm Parabellum ammunition which, by the late 1940s, had become the universal sub-machine gun round. It was a robust and well-made blowback gun, which was unusual in having a magazine and housing that could be unlocked and folded forward to lie underneath the barrel when being carried. It was quickly adopted by the French Army and saw widespread use in France's colonial wars in Indochina and Algeria.

The Australian Owen machine carbine was another wartime design that saw extensive combat service after World War II. Invented by a Lieutenant Owen in 1941 it went into immediate production for the hard-pressed Australian Army which needed a sub-machine gun for jungle fighting against the Japanese. Utilising a blowback system of operation the Owen was unusual in having a vertical top-mounted magazine which necessitated off-set sights. A rugged weapon, popular with combat troops, the Owen continued in service until the 1960s.

The Owen was eventually replaced by the F1, another Australian design intended to continue the Owen's reliability but which was of a lighter weight and would be easier to manufacture and maintain. The F1 has a straight-line stock which aids good shooting but calls for a high sight, which like the Owen is off-set because of the top-mounted magazine. This magazine is the same as that used with the British L2A3 Sterling sub-machine gun. The F1 was used by Australian troops in Vietnam, but some doubt was cast on its low-powered sub-machine gun round and it was dropped in favour of the American M16A1 assault rifle.

One feature which every designer has recognised as desirable in a sub-machine gun is small size, so as to make it convenient to carry and use. But the conventional blowback gun has its size set by certain physical constants: the body of the gun must be as long as the bolt and the compressed return spring behind it, otherwise the bolt will not have sufficient movement to load and extract. The size of the bolt is governed by the need to have a definite mass which will resist being blown open on firing, and with a given diameter or breadth of the gun body, this fixes the length of the bolt. Making the bolt too light means a too rapid opening of the breech and an over-fast rate of fire.

The first man to solve this dilemma was an Italian designer, Giovanni Oliani. In 1942 he broke new ground by producing a sub-machine gun which used an overhung or telescoping bolt. This development ensured that although the bolt had the necessary mass

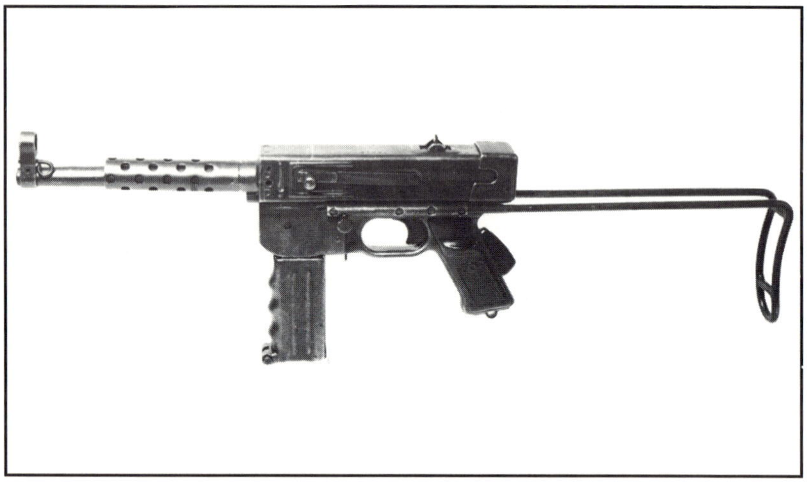

Above: The MAT-49.
Top: A Viet Cong guerrilla carrying a MAT-49 surrenders to US troops in Vietnam.
Below: The successor to the Owen, the F1 utilised the same top-loading magazine but had an improved stock which necessitated a raised rear sight.

to resist opening, the actual length behind the breech was no more than an inch or two instead of the six or eight inches of a conventional bolt. Thus the space behind the chamber could be reduced and with it the overall length of the weapon.

Oliani had little success with his overhung bolt and it was not until 1948 that his system appeared in military use, embodied in the Czech CZ23. The CZ23 was a sensible design with a tubular gun body and with the barrel set back into the body for some distance. The bolt was just over eight inches long but was hollowed out, so that when it was closed about six inches of it surrounded the rear of the barrel; instead of being overhung the bolt was telescoped over the barrel. Slots were cut in the bolt to allow the incoming cartridges to be fed into the chamber and the empty cases to be ejected; and the design was arranged so that the ejection ports in the bolt and in the gun body only coincided as the cartridge case was ready to be thrown clear. At all other times the two were out of register so that dirt could not enter the weapon through the ejection port. The design of the trigger mechanism was unusual: for single shots the trigger was pressed lightly while harder pressure produced automatic fire.

Besides bringing the telescoping hollow bolt into general use the CZ23 introduced a further though related innovation, namely the magazine being housed inside the pistol grip. The construction of a weapon with a hollow or overhung bolt brings the chamber well back into the gun body, so that the bolt mass can surround it, and if the pistol grip is pushed slightly forward from its usual position at the back of the body then it is ideally placed for adapting it as the magazine housing. There is a practical advantage in this: changing a sub-machine gun magazine in the dark can be a clumsy business when the magazine housing is 'somewhere out in front'. But if it is in the pistol grip, which is already being firmly grasped, then, acting on the principle that 'hand finds hand', it becomes easy to change magazines.

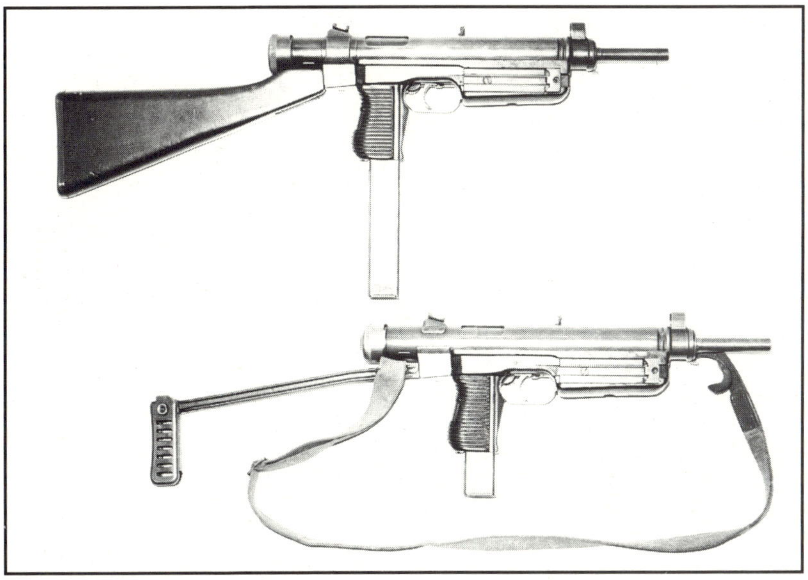

Above: The Czech CZ23 (top) with 40-round magazine, and the CZ25 with 24-round magazine. The CZ23 series was a major advance in sub-machine gun design as it made use of the telescoping hollow bolt; an innovation that was to be widely copied in future weapons.

The CZ23 was fitted with a wooden stock but was followed by the CZ25 model which had a folding steel stock; both types were taken into Czech service use. The CZ was chambered for the 9mm Parabellum cartridge, but in 1952 the Soviet Union insisted on the use of their standard 7.62mm sub-machine gun cartridge throughout the Warsaw Pact, so the CZ23 and 25 were withdrawn. They were replaced by the CZ24 and 26 which were exactly the same designs but chambered for the Russian cartridge. The 9mm models were then sold off to various countries, notably Syria and Cuba. In recent years the design has appeared again, in South Africa, as the Sanna 77. In this version it is restricted to semi-automatic fire only and was sold in the late 1970s as a home defence weapon for farmers. These Sanna 77s appear to be the original Czech-built guns refurbished and with the mechanism altered to prevent automatic fire.

Sub-machine guns

Type	Country	Calibre	Weight	Cyclic rate of fire	Muzzle velocity	Magazine
Thompson M1A1	USA	.45in	5.36kg (11.8lb) (loaded, 30 rounds)	700rpm	280mps (920fps)	20/30-round box 50-round drum
M3A1	USA	.45in	4.52kg (9.94lb) (loaded)	400rpm	280mps (920fps)	30-round box
Carl Gustav Model 45	Sweden	9mm	5kg (11lb) (loaded)	600rpm	360mps (1180fps)	36-round box
Madsen Model 50	Denmark	9mm	3.15kg (6.95lb) (empty)	550rpm	390mps (1280fps)	32-round box
MAT-49	France	9mm	4.76kg (10.59lb) (loaded)	600rpm	354mps (1160fps)	32-round box
MAS 38	France	7.65mm	2.85kg (6.3lb) (empty)	700rpm	350mps (1150fps)	32-round box
Owen	Australia	9mm	4.24kg (9.35lb) (empty)	700rpm	420mps (1375fps)	32-round box
F1	Australia	9mm	4.47kg (9.85lb) (loaded)	600rpm	360mps (1180fps)	34-round box
CZ25	Czechoslovakia	9mm	4.1kg (9.04lb) (loaded, 40 rounds)	650rpm	450mps (1475fps)	24/40-round box

Left: Besides equipping both regular and insurgent military forces the Uzi has proved a highly popular police weapon. Here a US presidential guard prepares to use his Uzi during the assassination attempt against President Reagan on 30 March 1981.

The Czech CZ23 was an influential sub-machine gun and its hollow bolt design was widely copied. One of the first – and certainly the most famous – copies was the Israeli Uzi. During the 1948 Arab-Israeli War the Israeli Army was poorly provided with sub-machine guns (most of what they had were old British Stens) and Israeli troops felt the lack of them in the close-quarter fighting encountered in the campaign. Shortly after the war had ended Major Uziel Gal set to work to examine existing sub-machine guns and design one for Israeli production. Working swiftly Gal produced a weapon – the Uzi – which entered service in 1951.

The most striking aspect of the Uzi was its small size and compact appearance, compared to most designs which preceded it, this being due to the adoption of the telescoping bolt and the magazine in the pistol grip. The Uzi bolt is square in section and hollowed out at its front end so that it wraps around the barrel, allowing the barrel to be set well back into the body. Construction is largely of steel pressings and heat-resistant plastics, and the tolerances are sufficiently liberal to allow the gun to go on working in combat conditions where dust and sand are prevalent. A press-in piece at the rear of the pistol grip acts as an effective safety device, only allowing the bolt to move forward if the weapon is being firmly held, so avoiding accidents if the weapon is dropped or allowed to fall on its butt.

Originally, the Uzi had a wooden stock but this was soon replaced by a folding steel one, and in 1960 the gun was further modified by the fitment of a larger bolt-retracting handle and an improved fire-selector catch. The Uzi has been widely admired for its compactness and reliability and it is licence-built in Belgium by FN who market the weapon in Europe, a

Uzi sub-machine gun

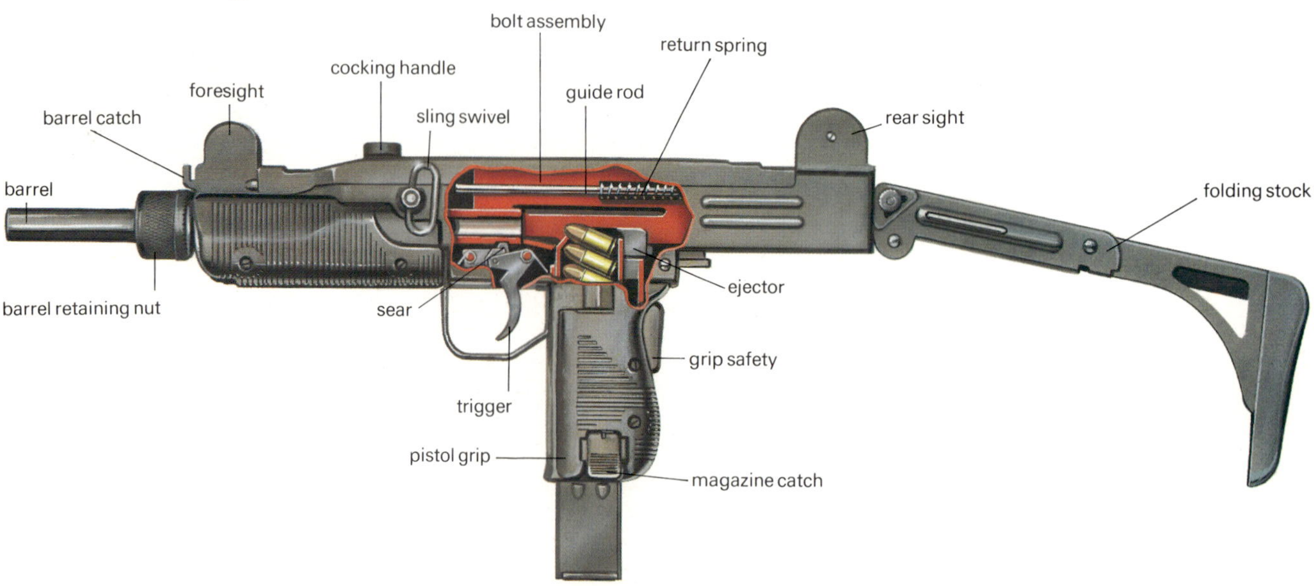

Although the Soviet Union has concentrated on the AK assault rifle, its World War II sub-machine guns are still in use today. Right: The PPSh-41, without magazine. Below right: The Type 50 is the Chinese copy of the PPSh-41 and comes with a 35-round box magazine (above), while the Chinese copy of the PPS-43 is the Type 54 (below).

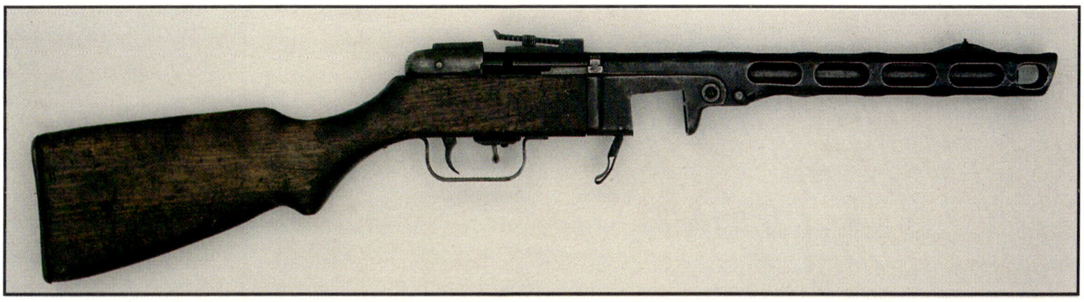

Italy has been an important manufacturer of sub-machine guns and even its older designs have had a long postwar career, notably the M38/49, depicted (below) with a member of the MPLA in the Angolan capital of Luanda during the independence celebrations of 1975. A newer type is the Model 12 which arms a number of states including Brazil's anti-terrorist police units (below right).

major customer being the West German Army where the Uzi is termed the MP-2. Outside Europe the Uzi has been exported to Iran, Venezuela and Thailand.

In Italy the Beretta Company had been making sub-machine guns since the end of World War I and during the 1930s they developed the Modello 38, a well-made and popular gun that eventually became the standard postwar sub-machine gun of the Italian Army under the designation M38/49. Designed to fire 9mm Parabellum ammunition the M38/49 was fitted with two triggers, the forward one firing single shots and the rear one full automatic fire. There are two sub-variants: the Model 4 fitted with a conventional manual safety catch and the Model 5 with an automatic catch.

During the 1950s Beretta began experimenting with a successor to the M38/49 and subsequently developed the Model 12. The Model 12 is completely built from metal pressings: the pistol grip, magazine

Above and above right: The Czech Skorpion sub-machine gun, with its small 7.65mm ammunition (centre) compared with standard 9mm (top) and a .303in rifle round. Below: The pistol-like Polish PM-63 with forward grip folded back. Below centre: The German MP40 of World War II fame. Bottom: The two modern West German Walther designs, the MP-L (left) and the short-barrelled MP-K (right).

housing, front grip and gun body are welded together to make one solid unit. Inside the body is the telescoped bolt, with about six inches of its length surrounding the barrel. There is a grip safety on the forward edge of the pistol grip which locks the bolt except when it is squeezed in by grasping the grip in the firing position. The Model 12 has been adopted by special units in the Italian Army and by a number of other armies, notably in the Middle East and Africa, and it has been licensed for manufacture in Brazil and Indonesia.

The Soviet Union, perhaps the greatest user of sub-machine guns during World War II, adopted the AK-47 assault rifle after the war as an all-purpose firearm and consequently had only limited interest in sub-machine gun development. Despite the AK-47, the vast number of sub-machine guns produced by the Soviet Union during the war ensured that they remained in use throughout the Warsaw Pact while others were exported to client nations. The most famous Soviet sub-machine gun was the PPSh-41 which was used throughout the war and became something of a symbol of the Red Army. Like many wartime designs the PPSh-41 utilised sheet metal pressing and extensive riveting and simple welding, although a bulky wooden stock gave it an old-fashioned pre-war look. A very reliable weapon, it worked by the standard blowback system and featured a 71-round drum magazine. Some five million PPSh-41 sub-machine guns were manufactured and post-1945 a large number of these were exported to the communists in China. After the communist takeover in 1949 the Chinese began manufacturing the PPSh themselves, calling it the Type 50. Even more rudimentary in construction than the Soviet original, the Type 50 was nevertheless an effective weapon, used extensively in Korea and later by the Viet Minh in Indochina.

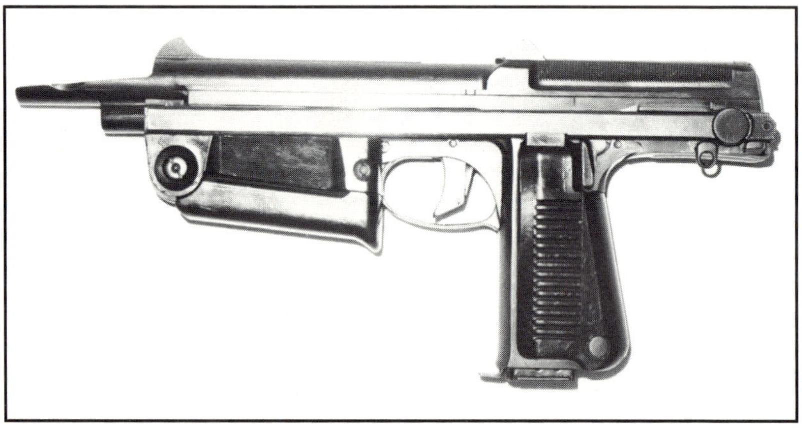

Another sub-machine gun of Soviet origin used in Asia was the PPS-43. Developed from the PPS-42 – a weapon brought into being during the siege of Leningrad – it was an even simpler sub-machine gun than the PPSh-41, constructed entirely from metal except for the plastic pistol grip. The metal shoulder stock could be folded over along the top of the receiver, making the PPS-43 a relatively compact weapon. Large numbers were sent to China after 1949, and in 1953 the Chinese instigated a mass-production programme for this weapon. Designated the Type 43 it has been widely used by China and her allies, including the North Vietnamese during the war against the Americans.

Apart from the Soviet Union the other important arms manufacturer in the Warsaw Pact is Czechoslovakia, which has developed a number of interesting sub-machine gun designs. After the famous CZ23 series, the major design has been the Vz61 Skorpion. The Skorpion operates on the usual blowback system utilising a solid bolt, but despite its conventional

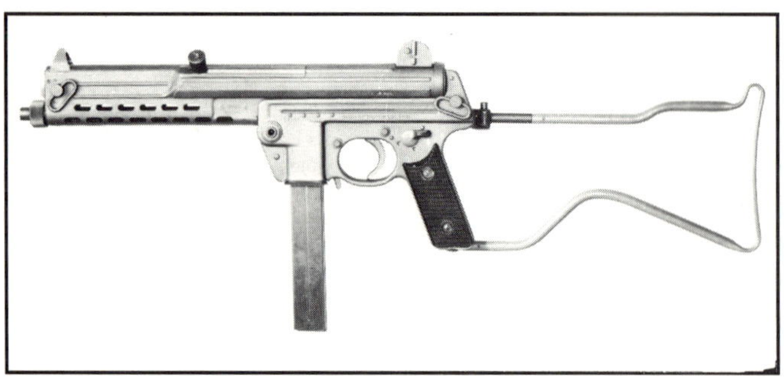

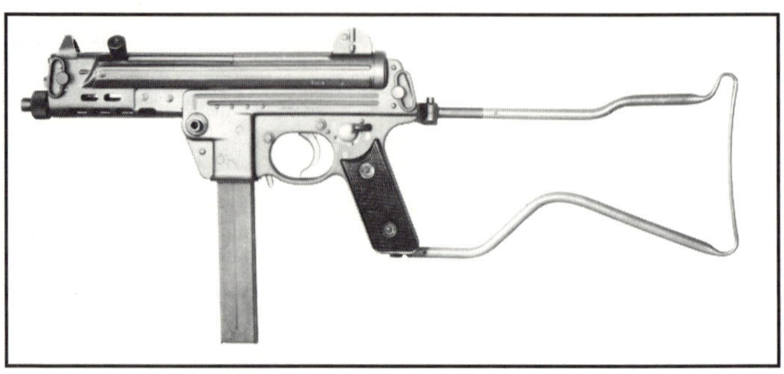

operation it is a highly unusual weapon. Instead of the 9mm or even the Soviet 7.62mm round the Skorpion was chambered for the 7.65mm automatic pistol cartridge, which is not generally considered a combat round and had never before been used in sub-machine guns. This led to a reduction in size, so that the Skorpion, with its wire butt folded, was no more than 27cm (10.6 inches) long. It is so small that it can be comfortably fired in one hand, like a pistol, if necessary. Indeed, the object behind it was to provide tank crews with a weapon they could wear in a holster and carry on their person but which had automatic fire power that could be used in the event of having to bail out from a wrecked tank in battle.

In theory the lightweight Skorpion – with a light bolt – would have had a very high rate of fire, but the designer foresaw this and fitted a very ingenious 'rate reducer'. Inside the pistol grip is a tube containing a weight and a spring. As the bolt recoils it strikes the weight and drives it down the tube, against the spring. The bolt is then held at the rear of the body, while the weight travels down and is thrown back by the spring; when it reappears at the top of the tube it trips a release and the bolt goes forward to fire the next shot. The delay thus induced is a mere fraction of a second, but it is sufficient to bring the rate of fire down to an acceptable 600 rounds per minute.

The Skorpion has since been made in larger sizes to take the 9mm short and 9mm Parabellum cartridges, but these are rarely seen. The original 7.65mm model is, however, a popular weapon with terrorists due to its small size making it easy to conceal.

Another small weapon, which, like the Skorpion is frequently classed as a 'machine pistol' rather than a sub-machine gun, is the Polish PM-63. Its design is really that of a large automatic pistol since it uses a moving slide instead of an internal bolt. The gun consists of a frame with pistol grip, into which goes a magazine and a fixed barrel. The slide surrounds the barrel, extends forward under the muzzle to act as a primitive compensator to stop the muzzle rising when firing automatic, and forms the breech block at the rear.

Once the weapon is cocked, the slide remains in the rear position, and it is released by pressing the trigger. It then runs forward, propelled by a spring, rams a cartridge into the breech and fires it. There is no form of breech lock and the slide is simply blown back, ejecting the empty case, compressing the return spring so as to be ready to reload and fire again. Left in this manner one would expect a high rate of fire, as with the Skorpion, but once again a rate reducer is utilised. The PM-63 is chambered for the Soviet 9mm Makarov cartridge, rather more powerful than the 7.65mm of the Skorpion but much less effective than the 9mm Parabellum.

The German MP40 sub-machine gun was a highly influential weapon; its use in the early campaigns of World War II demonstrated to the world the advantages of the sub-machine gun and led to a rush of attempts to develop easy-to-manufacture sub-machine guns, including the British Sten and the US M3. The MP40 was a development from the MP38 and the MP38/40, and its folding metal stock and the extensive use made of metal pressings set the precedent for future designs.

Since World War II the German arms manufacturers, Walther and Heckler and Koch have both produced their own sub-machine guns. The Walther

Heckler and Koch have brought out a wide range of sub-machine guns, all based around the MP5. Above: The original MP5 (here an export model) was initially designated the HK54 and was derived from the G3 rifle. Right: The MP5A1 is fitted with a sliding metal stock.

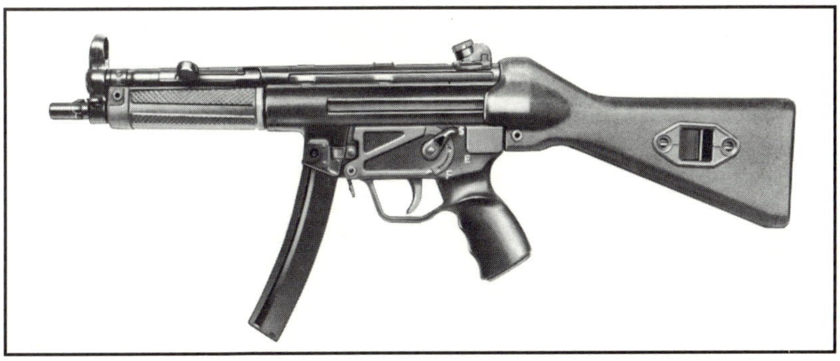

Above: The MP5A2 is distinguished from the standard MP5 by a new forward hand grip, a rotating rear sight and a modified front sight. Right: The MP5A3 is identical to the MP5A2 except that it is fitted with a sliding metal stock.

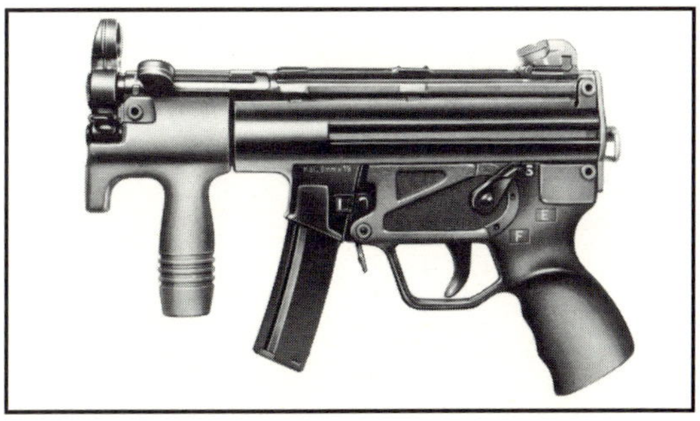

Above: The MP5 SD is a specially silenced model which can be fitted with the standard stock, with a sliding metal stock or with none at all. Right: The most unusual weapon in the MP5 range is the MP5K, used by anti-terrorist squads, which features a forward-mounted hand grip.

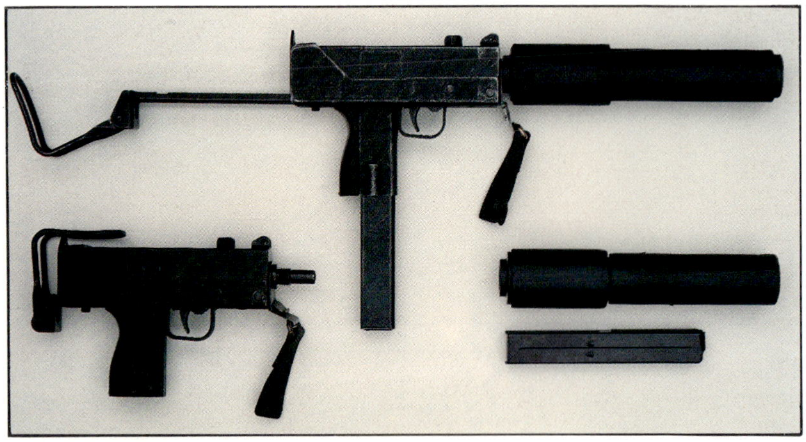

Above: The Ingram Model 10 (top) and Model 11 sub-machine guns have been privately designed as lightweight, easy to conceal weapons suitable for counter-terrorist operations. Apart from the barrel the whole weapon is made from metal pressings; even the bolt is manufactured from sheet metal and then filled with lead. Both models can be fitted with the sound suppressors illustrated.

MP-L/MP-K has a rather unusual shape since the barrel is below the tubular gun body. This is because the body serves to contain a large cylindrical section of the bolt, below which is a small appendage which actually loads and fires the cartridge. This allows much of the mass of the bolt to overhang the barrel at the moment of firing, a variation of the overhung bolt technique. The Walther is produced in two versions, the standard MP-L and the short-barrelled MP-K, and like most modern sub-machine guns has a folding stock. The Walther is widely used by European police forces and by some smaller armies.

The Heckler and Koch MP5 is unusual in that it is a delayed-blowback gun firing from a locked breech. The bolt is similar in some respects to the Walther design, with a large mass projecting over the barrel for compactness, but it is locked by two rollers engaging in recesses in the body, and these must be withdrawn from the recesses before the breech block can move backwards. The delay is extremely short, but this arrangement makes this weapon much more accurate than the vast majority of sub-machine guns.

A number of MP5 variants have been produced: the MP5A2 with a fixed stock; the MP5A3 with sliding metal stock; the MP5 SD silenced-version; the MP5K, a specially-developed counter-insurgency model which dispenses with a stock and has a shortened barrel with a second handgrip mounted forward of the magazine. The MP5 is favoured by, among others, the British SAS, and it is in military service with the West German, Swiss and Dutch armies as well as many police and security forces.

In spite of the success of the Thompson and M3 sub-machine guns, the United States military has shown little interest in the sub-machine gun since 1945. The only design to make much impact in America has been a private venture, promoted by designer Gordon Ingram. Following military service during World War II he returned to civil life intent upon making a sub-machine gun for police and military use. After several designs he developed the Model 10 which consisted of a square body made of steel pressings, an overhung bolt and a magazine in the pistol grip. The Model 10 featured a telescoping wire butt and when folded the gun was no more than 27cm (10.6in) long. The gun has a provision for a screwed bush around the barrel to allow a sound suppressor to be fitted. Ingram developed his gun in various calibres – the Model 10 which takes .45in and 9mm Parabellum ammunition, the Model 11 taking .38in – and it has been sold around the world, frequently being used by paramilitary and anti-terrorist units.

Ever since the end of World War II it has been predicted that the sub-machine gun will lose its military application, to be completely replaced by the assault rifle. This seems unlikely, but even if the military do abandon the sub-machine gun it still has a long career ahead of it as a weapon of guerrilla and counter-insurgency forces.

Sub-machine guns

Type	Country	Calibre	Weight	Cyclic rate of fire	Muzzle velocity	Magazine
Uzi	Israel	9mm	4.1kg (8.9lb) (loaded, 25 rounds)	600rpm	400mps (1312fps)	25/32/40-round box
M38/49	Italy	9mm	4.6kg (10.2lb) (loaded, 20 rounds)	500rpm	381mps (1250fps)	20/40-round box
Model 12	Italy	9mm	3kg (6.6lb) (empty)	550rpm	381mps (1250fps)	20/30/40-round box
PPSh-41	Soviet Union	7.62mm	5.3kg (11.7lb) (loaded, drum magazine)	900rpm	490mps (1600fps)	35-round box 71-round drum
PPS-43	Soviet Union	7.62mm	3.62kg (8lb) (loaded)	700rpm	490mps (1600fps)	35-round box
Vz-61	Czechoslovakia	7.65mm	1.55kg (3.4lb) (loaded, 20 rounds)	750rpm	294mps (970fps)	10/20-round box
PM-63	Poland	9mm	1.8kg (4lb) (loaded, 25 rounds)	600rpm	320mps 1050fps)	15/25/40-round box
MP-K	West Germany	9mm	3.43kg (6.6lb) (loaded)	550rpm	356mps (1168fps)	32-round box
MP5A3	West Germany	9mm	3.41kg (7.6lb) (loaded)	800rpm	400mps (1312fps)	15/30-round box
MP40	Germany	9mm	4.7kg (10.4lb) (loaded)	500rpm	381mps (1250fps)	32-round box
Ingram Model 10	USA	.45in	3.82kg (8.4lb) (loaded)	1100rpm	280mps (920fps)	30-round box

AK ASSAULT RIFLES

The Avtomat Kalashnikov, either as the AK-47, the later AKM, or the most recent AK-74, must be the most widely distributed rifle in history. It is the trademark of the Warsaw Pact and of their myriad satellites throughout the world, and such a wide adoption can only be due to general agreement on its excellence; even free gifts are discarded if they fail to function properly. The Kalashnikov has become a legend because of its functional appearance, its reliability and its simplicity. Not the most accurate rifle in the world, it is accurate enough for its purpose, and much of its success must be due to its 7.62mm short cartridge, an ideal combination of calibre, weight and power.

The man responsible for the famous Soviet AK assault rifle series was Mikhail Kalashnikov. Severely wounded as a tank commander during World War II, Kalashnikov was invalided out of the army and during his convalescence he put forward proposals for a sub-machine gun and a carbine, neither of which was accepted. Undeterred he continued working and in 1947 submitted drawings of a new rifle designed around a short 7.62mm cartridge which the Soviet Army had introduced at the end of the war. This time his design was accepted and it was introduced into Soviet service in 1953 as the *Avtomat Kalashnikov Obr 1947*, or AK-47. Since that time it has been estimated that something over 35 million AK series assault rifles have been made, and they are to be found in almost every country of the world; without any doubt the Kalashnikov is the most successful assault rifle ever designed.

What made the AK-47 such a success? Simply that it was sufficiently accurate, reliable, easy to make, easy to maintain and robust enough to stand up to the punishment of the battlefield. There is nothing revolutionary about the design, which is, in truth, a collection of features most of which have been seen on other weapons (most notably the German MP43/44 assault rifle). But its strength lies in the way these features have been brought together and welded into a near-perfect unity.

The AK-47 is a gas-actuated automatic weapon. Above the barrel lies a cylinder containing a piston; as the bullet is fired and passes up the barrel, a small portion of the propelling gas passes through a port and drives this piston backwards. At the rear end of the piston rod is a carrier which holds the bolt. As the rod and carrier move backwards, a shaped cam track in the carrier engages with a lug on the bolt and revolves

Page 211: Two Kalashnikov-armed Palestinian youths display their weapons for the benefit of a Western news photographer. The boy in the foreground has a conventional AKMS while his companion carries the Romanian-manufactured model with a pistol grip. Opposite page: The AK series was adopted by the Soviet Army in 1953 and has been used ever since, whether on active service in Czechoslovakia in 1968 (below) or on manoeuvres in the Soviet Union in 1981 (above), in which naval infantry are shown advancing in winter conditions supported by MICVs. Above: Troops of the Afghan Army parade in Kabul in 1979 armed with AKMs. Right: A Soviet colour guard of sailors pose in typical fashion. Of interest is the difference between the magazines: the one on the left is made from stamped steel and the other from plastic.

the bolt, unlocking it from engagement with the rifle breech. Once unlocked the rearward movement of the carrier withdraws the bolt, pulls the empty cartridge case from the chamber and ejects it. At the same time a return spring is compressed, and the firing hammer is cocked by the movement of the carrier. Then the spring forces the carrier and rod back, collecting a fresh cartridge from the magazine and loading it into the chamber. The bolt stops, but the carrier continues forward so that the cam now rotates the bolt in the opposite direction and locks it into the rifle barrel ready for firing. As the soldier pulls the trigger, the hammer springs up and hits the firing pin in the centre of the bolt, and the cartridge is fired, to begin the process over again.

By moving a lever on the right-hand side of the rifle, the soldier can select automatic fire, and in this case the rifle will continue to load, fire, eject and reload so long as the trigger is held down, firing off rounds at a cyclic rate of about 600 rounds per minute, though since the magazine holds only 30 rounds this rate obviously cannot be achieved in practice. In fact, with the fire selector set on 'auto', a rate of 90 rounds per minute would be considered acceptable.

A particular feature of assault rifles – of which the AK-47 is no exception – is the use of a smaller

intermediate cartridge. The M43 7.62mmx39 intermediate round of the AK-47 is lighter than the standard rifle cartridge, and while less powerful it is still capable of carrying out the function assigned to it. In addition, the intermediate round is much more powerful and more accurate than the pistol rounds used in sub-machine guns. The M43 round has a higher trajectory than almost any other ammunition type and as a consequence is ineffective at ranges over 400m (440yds). However, since it is accepted that few soldiers are capable of hitting anything over 200m (220yds), and that most fire-fights occur at much shorter ranges, this lack of long-range effectiveness is not particularly damaging to the rifle's overall performance.

Despite its ruggedness and reliability of operation the AK series does have a few faults, most notably the vulnerability of the rather exposed gas tube, which if damaged will cause operational malfunctions brought about by uneven gas pressure. Also, the rifle has a tendency to overheat during sustained fire, making it difficult to handle and more seriously bringing about the danger of the round spontaneously 'cooking-off' in the chamber.

The AK-47 was manufactured in the traditional manner, the receiver being machined from a block of steel. By modern standards this is uneconomic, and in 1959 a 'modernised' version, the AKM, was produced. This uses stamped steel for the receiver and riveting for the assembly, but otherwise is the same as the earlier model. Both the AK-47 and the AKM can

Above: Moslems brandish Chinese- and Soviet-made weapons during a lull in the fighting for Beirut during the civil war in Lebanon in 1975. Besides the AK rifles (with folding bayonets attached) are an RPD light machine gun (with bipod) and an RPG7 rocket launcher. Left: A Lebanese militiaman inspects the ground below his position. Note the religious picture stuck onto the stock of his AKM, a traditional custom among Christian troops in Lebanon. Right: This Israeli soldier follows the practice, made popular in Vietnam, of taping an extra magazine (upside down) to the magazine in the rifle, so that during combat the second magazine can be quickly flipped round, in effect doubling the immediately available ammunition supply.

be found with wooden butts or with folding steel stocks. Motorised and airborne units use the lighter folding stock version which is termed the AKMS. The AKM can be fitted with a NSP-2 infra-red night sight and a knife bayonet that can also function – unlike that of the AK-47 – as an insulated wire cutter and a miniature saw. Additionally, a grenade-launcher attachment can be fitted to the basic model.

As well as outfitting the Soviet Army the Kalashnikov design has been exported to the communist bloc and to a large number of other countries, several of which have begun manufacture themselves, sometimes incorporating minor variations. The Chinese designate it the Type 56 and fit a folding bayonet to the muzzle; the Romanians fit a wooden front pistol grip so that it can be used as a sub-machine gun, while the Hungarians use a perforated metal fore-end and a nylon front grip. Butts of natural wood, laminated plywood and plastic materials are employed by all these nations.

In addition, there are 'Kalashnikov look-alikes'. The Yugoslavian Army is equipped with the M70 rifle, a close copy of the Kalashnikov, and Yugoslavia also produces the M77 which is in 7.62mm Nato calibre for export. The Finns have the Valmet M62/M70; this began as a straight copy of the AK-47 and has since been modified, though in appearance its AK-47 ancestry is quite apparent.

In 1978 the West became aware of a new version, the AK-74, which is essentially an AKM but with a smaller 5.45mm calibre. This development is in line with the adoption of the 5.56mm cartridge by the Americans for their successful M16 rifle. The smaller calibre ensures a lighter cartridge with less recoil force so that automatic fire from the shoulder is more easily controlled and more accurate. While being lighter than the M43 round, the magazine – made of plastic – still holds the standard 30 rounds. The most prominent recognition feature of the AK-74 is the muzzle brake and compensator which directs some of the emerging gases up and to the right to counter the rifle's tendency to climb when fired on automatic. It also reduces the recoil force, making the weapon easier to shoot. A further advantage over the 7.62mm version is the AK-74's increased muzzle velocity (900 metres per second as against 715 metres per second) which will provide a flatter trajectory and so greater accuracy and impact force.

The AK series design has also extended to acting in

Top: The AK series has been one of the most exported rifles of the postwar period. This Yemeni guerrilla proudly displays his AK-47, complete with home-made leather casing. Above: Palestinian commandos in Lebanon in 1982. The soldier in the foreground is armed with an AKMS fitted with a rocket-launcher attachment.

AKM assault rifle

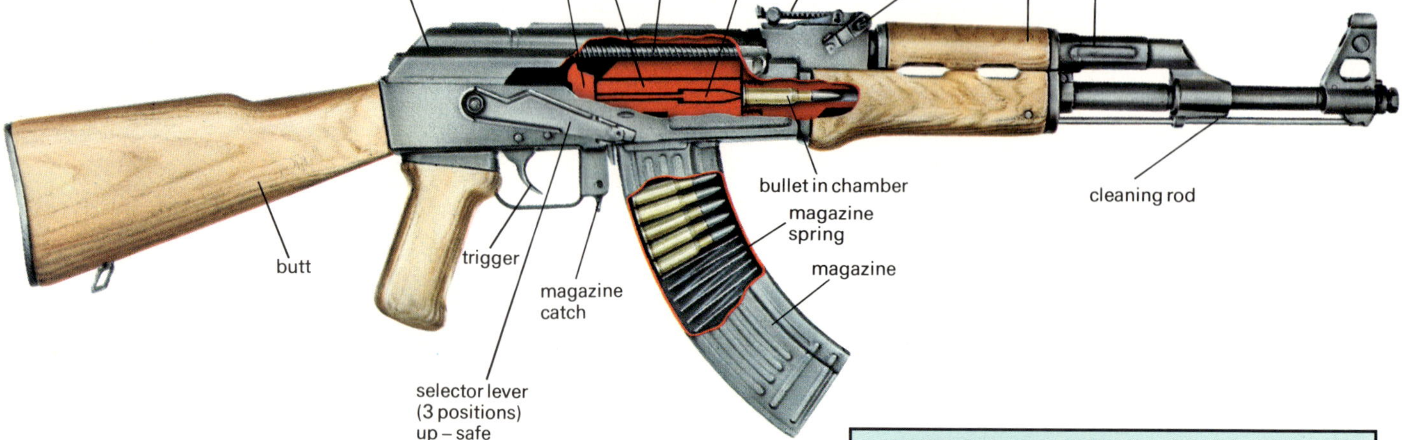

AKM Assault Rifle

Calibre 7.62mm
Length (without bayonet) 88.8cm (35in); AKMS 65.3cm (25.7in) with stock folded
Weight (loaded) 3.80kg (8.38lb); AKMS 3.46kg (7.62lb)
Rate of fire Cyclic 600rpm; practical auto 90rpm; practical semi-auto 40 rpm
Maximum sighted range 1000m (1100yds)
Effective range Auto 200m (220yds); semi-auto 300m (330yds)
Ammunition Ball, tracer, armour piercing, blank, grenade launching
Magazine 30-round box
Cartridge M43 7.62mmx39 intermediate round
Muzzle velocity 715mps (2345fps)

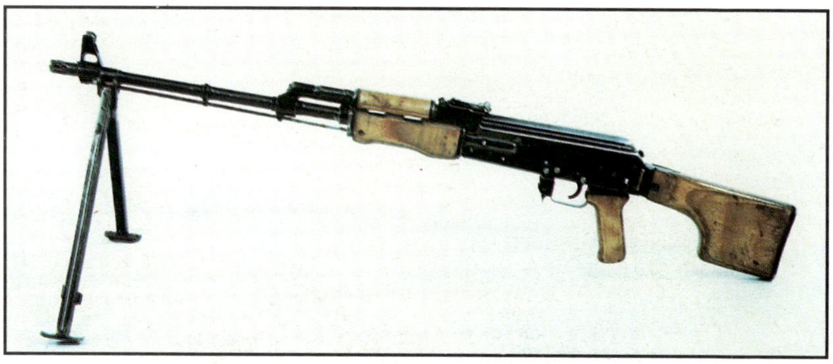

Above left: An RPK light machine gun photographed without its magazine. Left: The AK-74, which will eventually replace the AKM in the Soviet armed forces. Firing a 5.45mm calibre bullet the AK-74 follows the trend set in the West for small-calibre rifles. Below left: Armed with AKMs and an RPD LMG, Afghan guerrillas engage Soviet troops.

a light machine gun role, namely the RPK and the PK models. Introduced in 1960 the RPK is essentially a bipod-mounted AKM but with a longer, stronger barrel, a different stock and a larger capacity magazine – either a 40-round box or a 75-round drum magazine. Light and very handy to use the RPK is, however, too akin to the AKM to be really effective as a light machine gun: its M43 cartridge is not powerful enough for long range fire and its (unchangeable) barrel overheats rapidly. These problems were largely overcome with the introduction of the PK model which, weighing nearly twice as much as the RPK, fires the full power 7.62mm cartridge and, belt fed, is capable of sustained long-range fire. Highly regarded in the West – following tests by the US Army – this light machine gun can be either bipod or tripod mounted; and in the latter firing position it is termed the PKMB.

FN FAL ASSAULT RIFLE

The Fabrique Nationale 'Fusil Automatique Léger' is probably, next to the Soviet AK series, the most widely used modern rifle. Developed to take advantage of the Nato 7.62mm cartridge, its appeal spread far beyond the confines of Nato, and, at the most recent count, it has been adopted in over 90 countries. There have been several variant models, ranging from 'paratroop' versions with folding butt to heavy-barrelled versions which take over the role of the light machine gun. The simple mechanism has remained the same, however, giving a legendary degree of reliability.

Fabrique Nationale Herstal SA, the well-known Belgian gunmakers, began experimenting with automatic military rifles well before World War II, the idea being to replace the conventional bolt-action rifle with a gas-operated model. The German invasion of Belgium in 1940 put a temporary halt to this programme but after the end of hostilities a new rifle was designed and was put into production as the ABL (Arme Belgique Légère). It was supplied to the Belgian, Venezuelan, Egyptian and other armies in various calibres.

As the ABL began its manufacturing run, FN then began contemplating the idea of producing an assault rifle of modern type, and built a prototype using the wartime German 7.92mm 'Kurz' cartridge. By the late 1940s the British were busy developing weapons for their experimental .280 (7mm) round in the hope of seeing it adopted by Nato, and FN rebuilt their new rifle to suit this calibre. But although the .280 was a good cartridge, the Americans refused to accept it, at that time having little interest in the 'assault rifle' concept of short-range light weapons. FN realised that the Americans – as the senior partner in Nato – were going to push their .30 T65 cartridge into acceptance, and so they immediately began redesigning their assault rifle for the new round. When the American T65 was formally approved as the 7.62 × 51mm Nato cartridge, FN were ready with a working rifle, the FAL or Fusil Automatique Léger.

The FAL is one of the few military automatic rifles in use today which does not use a rotating bolt. It is gas-operated, and a portion of the gas following the bullet is vented through a port in the barrel to drive a gas piston backwards. The rear end of the piston rod strikes the bolt carrier, driving it backwards to compress a spring. Below and enclosed by the carrier, lies the bolt, its rear end pressed down in front of a transom in the rifle body which prevents it moving back during firing. As the bolt carrier moves back, so shaped cams on its side strike two lugs at the sides of the bolt and lift the rear end out of engagement with the transom. The bolt is then free to be pulled back by the carrier, extracting the spent case and ejecting it. The rearward movement also passes over the hammer and cocks it

Above and right: Both British and Argentinian troops were armed with FN rifles during the Falklands conflict. The Argentinian soldiers depicted here carry the post-1964 pattern rifle, very similar in appearance to the British-produced SLR (self-loading rifle).

Previous page: Few firearms have seen such widespread service as the FN FAL, and it was a popular weapon during the fighting in the former Belgian colony of the Congo, here being carried by these Katangan mercenaries.

Right: Armed with SLRs a patrol from the Rhodesian security forces carefully makes its way through the bush as part of a search and destroy mission against Patriotic Front guerrillas.

Right: A highly dramatic firing position is adopted by this SLR-armed trooper of the Rhodesian Greys Scouts. More a pose for the camera than anything else, an aimed shot of this type would have little chance of hitting anything smaller than the side of a barn. Below right: The other extreme – taking few chances FNLA guerrillas gun down a civilian at short range during fighting in Angola in 1975.

ready for the next shot. The spring then drives the carrier and bolt forward; the bolt face collects a fresh cartridge from the magazine and forces it into the chamber, and as the bolt carrier runs forward so the cams now force the rear end down into engagement with the transom once more. The firer pulls the trigger to release the hammer and the next shot is fired.

The gas is channelled to the gas piston through a regulator which is adjustable. The object of this is to ensure that there is always sufficient gas to operate the mechanism, even though it may be stiff from dust or lack of lubrication after prolonged firing. Adjustment of the early design of regulator was a long and tiresome business but a more simple design was later adopted. In addition, since the rifle can be used for launching grenades (by slipping their hollow tails over the rifle muzzle), the regulator permits the supply of gas to the piston to be completely shut off so that all the available gas is used to propel the grenade. When launching grenades a special blank cartridge is usually provided, though some new grenade designs permit the use of ordinary ball ammunition for this purpose. A number of sights can be fitted; besides the sniper's telescopic sight, the FAL can take the Trilux night sight. In service with the British Army since 1974, and designated SUIT (Sight Unit Infantry Trilux), it is self-energising and can be used to observe indistinct targets in daylight as well as at night.

Since the British 7mm rifle could not be redesigned to suit the 7.62mm cartridge, the British Army were among FNs first customers for the new weapon, adopting it as the SLR L1A1; the SLR's first service use was against the Mau Mau in Kenya in 1954. Production was begun in the Royal Small Arms

Factory, leaving the FN plant in Belgium available for other orders, which came in profusion. Licences to manufacture also went to Argentina, Australia, Austria, Canada, India, South Africa and Norway, while exports were made to almost every country outside the communist bloc. A total of 90 countries have used the FN FAL as their standard rifle. Much of its popularity can be attributed to its robust design and overall reliability. This makes it virtually 'soldier proof', but at the same time gives the infantryman a reasonably accurate weapon with good stopping power.

As might be expected, not every army wanted the same specification, and therefore there are innumerable minor variations of the FN FAL in existence. The British L1A1, for example, fires only single shots and has had all its dimensions changed to suit British measurements and manufacturing techniques; it also has deep oblique cuts in the bolt carrier which collect dust and dirt that might otherwise jam the mechanism in battle conditions. The Canadian C1 version has a unique five-position sight and a two-piece firing pin; the Australian F1 model has a shorter barrel and butt than usual; the Netherlands Army has modified its FN to take a fixed 150m (165yd) sight, a metal handguard and a permanently-attached bipod. Although the standard FAL is capable of firing on automatic, problems of accuracy arose when long bursts were

Opposite page: British troops have employed the L1A1 SLR since the 1950s, both in training (above) and in earnest on the streets of Northern Ireland (below). Right: A member of the IRA takes aim with his FN; a possible target might be the similarly-armed British soldier at the scene of an incident in Northern Ireland (below). Bottom: Israeli soldiers set about cleaning their FN rifles during a break in fighting in the Negev Desert.

fired as the light-weight barrel tended to rise above the target. As a result, many countries followed the British and had the rifle fixed on semi-automatic, which nevertheless allows 30 well-aimed shots to be fired every minute. By contrast, FN has produced a number of heavy-barrelled models which, with bipods and full automatic fire capability, are used as squad light machine guns by some armies. Short-barrelled carbine versions are also made, as are models with folding butts for parachute troops.

The FN FAL is now reaching the end of its service life with several major armies. The British have indicated that they will soon go over to their own design in 5.56mm calibre; the Austrians are replacing it with their 5.56mm Steyr AUG, the South Africans with the R4, a locally-produced version of the 5.56mm Israeli Galil. As early as 1963 FN debated the possibility of scaling-down the FN to 5.56mm calibre so as to be ready for the new trend, but adopted a different method of bolt operation for their CAL – Carabine Automatique Légère – model. A particular feature of the CAL was its ability to fire three-shot bursts with one pull of the trigger, as well as operating on the standard automatic and semi-automatic modes. In the event the CAL was ahead of its time, largely because when it was introduced in 1966 few of the world's armies were ready to make the change to 5.56mm calibre. Subsequently the design was dropped in favour of another known as the FNC (Fabrique Nationale Carabine) which is now being considered by various armies as a possible replacement for their FAL models.

L1A1 SLR

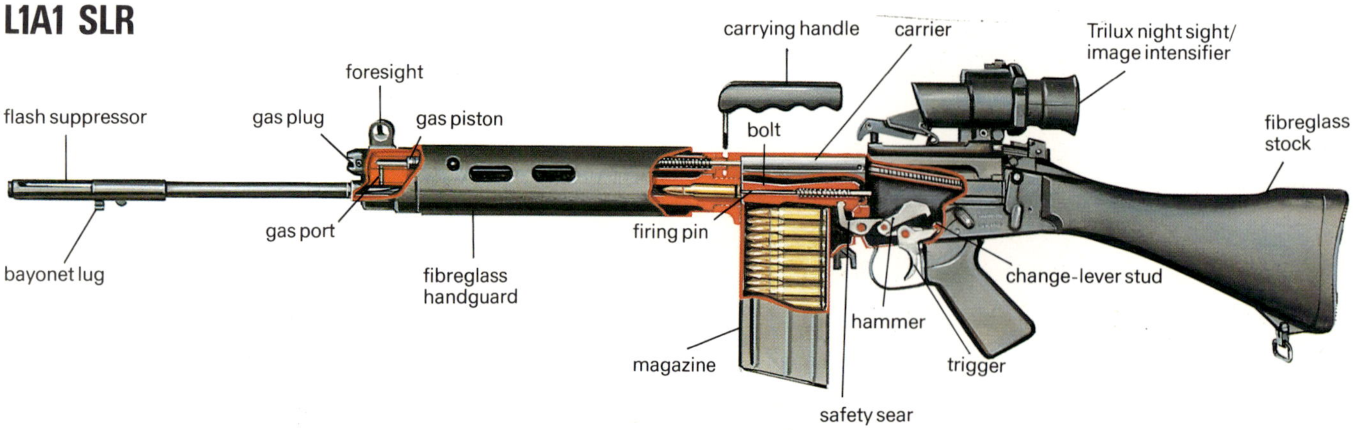

FN FAL

Calibre 7.62mm
Length 105cm (41.5in)
Weight (loaded) 4.31kg (9.5lb)
Rate of fire Cyclic 650-700rpm; practical semi-automatic 30rpm
Maximum sighted range 600m (650yds)
Ammunition Ball, tracer, armour piercing, blank, grenade launching
Magazine 20-round box
Cartridge 7.62 x 51mm Nato round
Muzzle velocity 853mps (2800fps)

Right: The successor to the FN FAL was the 5.56mm FNC, available with either the standard barrel as the Model 2000 (above right) or with the shorter barrel as the Model 7000 (right and far right). Both models have fold-away stocks.

Right: The FNC is a leader in the trend towards the basic rifle being the central element of a complete weapons system. Below the rifle is a clip-on bipod, a magazine (interchangeable with the M16A1 magazine) and a compact cleaning kit which slots into the pistol grip. By the muzzle is the blank round firing attachment, and directly above that are an M7 and a standard tubular bayonet, surmounted by a rifle grenade. Two telescopic sights are provided; the smaller one, the basic infantryman's scope and the larger, a special sniper's model. On the rifle itself the change lever has four possible positions: S – safety; 1 – single shot/ semi-automatic; 3 – controlled burst of three rounds; A – fully automatic.

G3 ASSAULT RIFLE

The Heckler and Koch G3 rifle, another design which is in service across the world, has its origins in the research establishments of Germany during World War II. From a basic idea in 1944 it was gradually improved and refined until it became the basis for a whole family of weapons ranging from sub-machine guns to machine guns. The G3 is something of an oddity in today's armoury since it is almost the only major rifle design which relies on the delayed blowback principle.

The West German Gewehr-3 (G3) assault rifle presently equips the armed forces of some 40 countries throughout the world, from South America and the Middle East to Africa and Southeast Asia. Its popularity as a weapon, especially in those countries with limited defence budgets and no arms-manufacturing facilities, results from its position as the centre pin of a versatile family of weapons designed to fulfil a variety of combat roles. The weapons-family concept has the advantages of drastically reducing the complexity of training involved in the use of infantry company weapons, from individual smallarm to unit-support machine gun, and simplifying the logistical process of supplying units in the field. A great many of the components are interchangeable between the weapons designed for the various roles, easing the pressure on parts replacement, especially on the battlefield. A further major attraction of the system is that it is manufactured in three different calibres to accommodate not only the standard Nato 7.62 × 51mm round, but also the lighter intermediate 5.56 × 45mm and the Soviet 7.62 × 39mm cartridges; this can be an important factor when potential clients are considering future sources and availability of ammunition.

The initial concept and early development of what was to become the G3 took place in Germany towards the end of World War II at the Mauser works at Oberndorf am Neckar. The product of this research was the StG 45M assault rifle which featured an unusual delayed-blowback mechanism based on a two-piece bolt with a twin-roller locking mechanism. It weighed 3.7kg (8.18lb) and was fed from a 30-round detachable box magazine. Constructed from metal stampings, the StG 45M was a simple, well-balanced weapon but the design was never fully developed before the close of the war. Research along the same lines, however, continued. After the war, Ludwig Volgrimmler, one of the original Mauser engineers who had worked on the StG 45M, went first to the French arms centre at Mulhouse, where he designed two further delayed-blowback breech mechanisms for the US 7.62 × 33mm carbine cartridge, and then on to the Centro de Estudios Técnicos de Materiales Especiales (CETME), the Spanish government weapons research establishment. At CETME Volgrimmler, with a number of other German and Spanish designers, produced the prototype of the CETME automatic rifle which was ready for trials in 1952. The original CETME rifle was chambered for a special CETME short-cased lightweight round, designed for high velocity, but further tests led to the adoption of a reduced-power version of the standard Nato 7.62mm round. Research continued and, in 1954, the German firm of Heckler and Koch was assigned the licence to develop the CETME rifle. Over the next three years they modified the weapon into the G3, capable of firing the full-power Nato cartridge. By 1959 the G3 had been accepted as the Bundeswehr standard assault rifle, replacing the German G1-designated Belgian FN FAL, and since then a range of weapons has appeared from Heckler and Koch based on the G3's delayed-blowback mechan-

Previous page: With his G3A3 at the ready, a nervous El Salvadorian government militiaman stands point at a street corner while (above) surrounded by spent cartridge cases, an El Salvadorian regular engages guerrillas with his G3 on sustained fire.

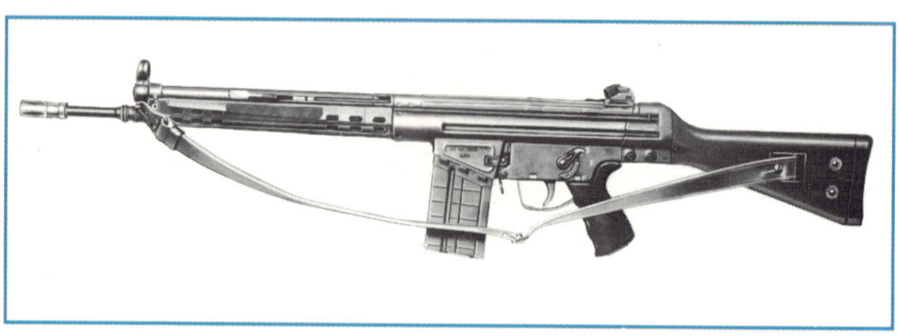

Above: The Spanish 7.62mm CETME assault rifle which features a roller-delay locking mechanism and was the forerunner for the design of the Heckler and Koch G3 series. Below: The standard Heckler and Koch G3A3 assault rifle, issued to the German Bundeswehr.

ism. The current standard G3 model is the G3A3, although there are several other G3 marks which incorporate minor modifications. The G3A1 is fitted with a folding stock and flip-over backsight, the G3A2 has a rotating backsight, and the G3A4, designed for use by paratroopers and AFV crews, has a retractable stock to provide a more compact version of the basic weapon. A sniping version, the G3A3ZF, is also available.

One of the basic design problems of all blowback, as opposed to gas-operated, weapons is that at the moment of firing of a high-power cartridge, the internal pressure forces the cartridge case hard against the walls of the chamber. Unless some kind of delay mechanism is built in, the case head can be torn off as the mechanism tries to eject the spent case and the weapon will jam. The G3 overcomes this problem by means of a two-piece bolt and the provision of locking rollers in the barrel extension. When the trigger is pulled, the hammer strikes the firing pin which fires the cartridge. The gases released by the firing drive the bullet down the barrel and the locking rollers, which have been forced outwards into the recesses in the barrel extension by the forward movement of the locking piece and bolt head carrier, prevent the immediate unrestricted recoiling of the bolt. As the expanding gases force the bolt head back, pressure is applied to the locking rollers which in turn force the locking piece and bolt head carrier backwards. The bolt head and spent cartridge case are now free to recoil and the case is ejected. In effect, this arrangement keeps the breech closed until the bullet leaves the muzzle and the pressure drops to a safe level before ejection takes place. Initial extraction of the cartridge case is aided by 12 flutes cut into the chamber which allow the propellant gases to leak forwards along the cartridge and float the mouth of the case away from the front of the chamber. The recoiling of the bolt then compresses the return spring which drives the bolt forward to repeat the cycle.

Most automatic rifles, such as the M16, operate by a gas port and piston system, but on the G3 neither is necessary and problems with blockage are eliminated. The G3 system provides a cyclic rate of fire of between 500 and 600 rounds per minute and is said to be capable of accepting virtually any brand of ammunition supplied by Nato countries or acquired elsewhere in time of war.

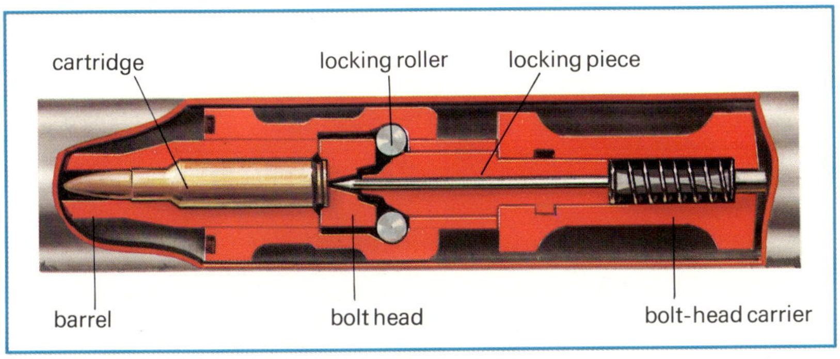

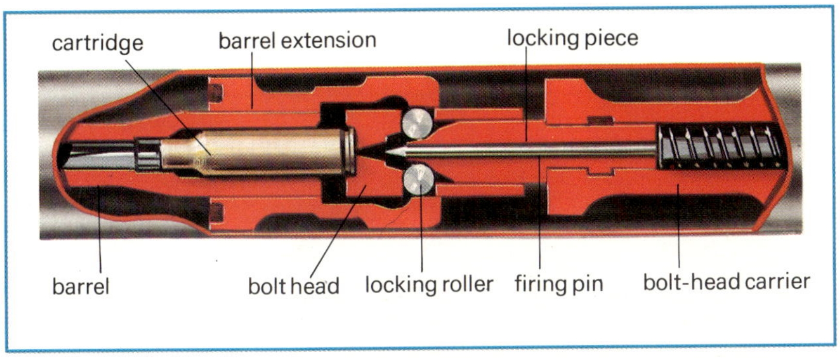

Above (top to bottom): An export version of the G3, field stripped to show its main components; the G3AZF sniper version; two diagrams showing the G3's delay mechanism. In the top diagram the bolt is locked while below, after the bullet has left the muzzle, the bolt is free to recoil. Right: The G3 has seen action in a number of African countries with both colonial and guerrilla forces, and can be seen here as part of a motley selection of smallarms fielded by a ragged unit of FNLA guerrillas in Angola.

The G3 is constructed mainly from metal stampings, making it a cheap and easy weapon to manufacture. Like the M16 and the FN FAL, it is fitted with a tough plastic stock and forward handguard. It is easy to strip down and maintain in the field and can be fitted with a range of accessories. Rifle grenades can be fired without the addition of a muzzle attachment and the standard flash suppressor doubles as a guide for the fitting of grenades. An infra-red night sight is also available. For training purposes, the G3 can be fitted with a blank-firing adaptor for both semi-automatic and automatic fire.

In the 7.62 × 51mm calibre group, two further weapon variants are available which are closely related to the parent G3A3. The HK11 light machine gun is constructed from the same group of components as the assault rifle and can be fed by either a 20-round box or an 80-round double-drum magazine. A front-mounted bipod is available for use with the box, while a centre-mounted bipod is fitted when using the larger drum magazine. Similar in many ways to the HK11 is the HK21 machine gun. The HK21, however, is considerably heavier and is usually fed from a metal-link belt, although it can, by means of a special adaptor, accommodate a box magazine. Designed specially for sustained fire, the HK21 has a cyclic rate of fire of some 850-900 rounds per minute and is equipped with a quick-change barrel. A bipod can be fitted in both the front and centre positions, as on the HK11, but for heavy sustained usage a tripod mounting is also available. A Swedish variant of the HK21, the HK21A, differs in that the feed tray is hinged rather than removeable.

In 1963 Heckler and Koch began work on a new, scaled-down version of the G3, chambered for the less powerful 5.56 × 45mm cartridge. After five years, and a number of modifications, it was put into production as the HK33 assault rifle. While maintaining the same method of operation as the G3, the HK33 weighed less, and the lower-powered intermediate

Above top: The HK11 light machine gun version of the G3, fitted with a bipod and 30-round box magazine. Above centre: The 7.62mm HK21A1 metal-link belt-fed machine gun. Above: The scaled-down version of the G3, the HK33, which is chambered for the 5.56 × 45mm intermediate round.

Left: A unit of Portuguese infantry awaits further orders during a search and destroy operation in Mozambique. The G3 is the standard infantry weapon of the Portuguese armed forces and the soldier in the foreground has the belt-fed machine gun version, the HK21.

round considerably improved its accuracy on automatic fire. The HK33 also had a much higher muzzle velocity than the G3 and a slightly improved cyclic rate of fire. Feed is provided by a 20-, 25-, 30- or 40-round box magazine and the weapon can be fitted with bayonet, blank-firing attachment, rifle grenades, bipod and telescopic sight. Variants on the HK33 include the HK33K, a carbine version with shorter barrel and retractable stock, and the HK53, also a reduced-barrel version with retractable stock, designed for use in the sub-machine gun role. In keeping with the Heckler and Koch weapon-family concept, the HK33 is also manufactured as the HK13 light machine gun, fed by a 100-round dual-drum or 25-round box magazine. The 5.56 × 45mm group further includes the HK21 machine gun version and the HK23, a lighter version of the HK21A.

With an eye open to a market in those nations which receive smallarms from the Soviet Union, Heckler and Koch have also manufactured the G3-based series as the HK32 assault rifle, the HK12 light machine gun and the HK21 machine gun, all chambered for the Soviet M1943 7.62 × 39mm cartridge. The HK32 is fed from a 30-round box magazine but, although identical in dimension and operation to the HK33, it has a considerably reduced muzzle velocity and degree of accuracy. An HK32K carbine version has also been produced.

Since the early 1950s, the original concept and research behind the delayed-blowback mechanism has been thoroughly exploited with a view to supplying armed forces the world over. Heckler and Koch continuously update their systems and have produced A1, A2 and A3 versions. In the early 1980s, a number of E-designated variants appeared featuring a 3-round burst facility in addition to the single shot and continuous fire modes. Apart from German manufacture, the G3 has been built under licence in Brazil, Pakistan, Portugal, Iran, Norway, Sweden and Spain. Manufacture has also been subcontracted to both the French SAV-CIE arms factory and the British Royal Small Arms Factory at Enfield. It has seen action in a number of countries including El Salvador and Iran and was put to extensive use by the Portuguese in Angola, Mozambique and Guinea. Its versatility as a family of weapons is undeniable, especially when used by armed forces lacking the high standard of training demanded by many modern and sophisticated weapon systems, and it is ideally suited to the conditions of small-unit action.

Below: Like the G3, the HK33 is also manufactured in a retractable-stock carbine version, the HK33K. The weapon illustrated is an HK33KA1. Centre: Another shortened version of the HK33 is the 5.56mm HK53 sub-machine gun which weighs only 3.05kg (6.72lb) unloaded.

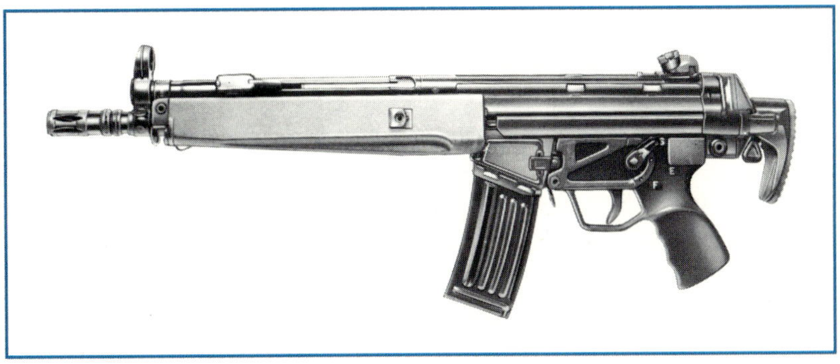

Left: A black South African infantryman stationed near the border with Angola, proudly shows off his G3A3 for the photographer.

Right: Keeping a tight hold on his G3 rifle and an eye open for any signs of enemy movement, an El Salvadorian soldier crawls through undergrowth during an anti-guerrilla operation in El Salvador's countryside.

G3 Assault Rifle

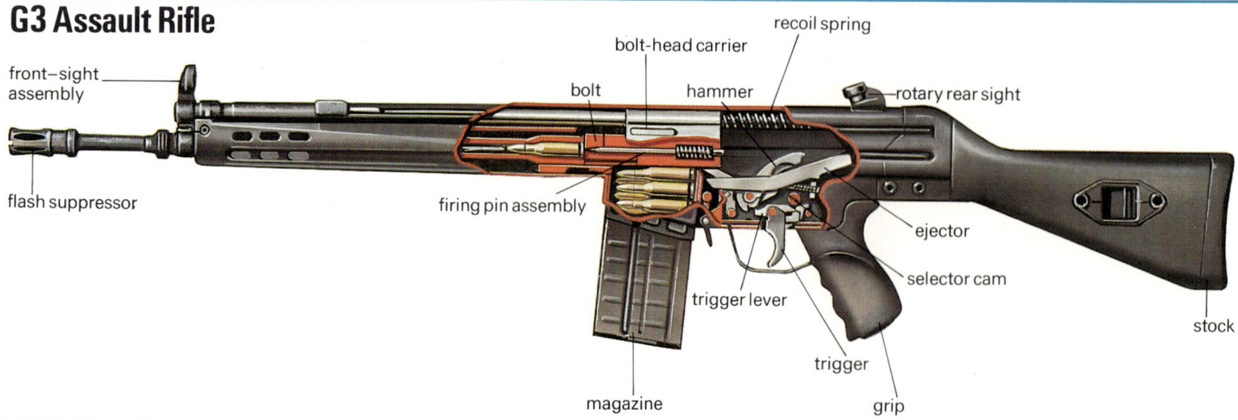

G3A3 Assault Rifle

Calibre 7.62mm
Length (fixed stock) 102.5cm (40.38in)
Weight (loaded with 20-round box magazine) 5kg (11lb)
Rate of fire Cyclic 500-600rpm; practical automatic 100rpm; practical single-shot 40rpm
Maximum effective range 400m (440yds)
Magazine 20-round box
Cartridge 7.62 × 51mm Nato
Muzzle velocity 780-800mps (2458-2624fps)

Above: The 5.56 × 45mm calibre group of Heckler and Koch infantry weapons also includes the HK13 light machine gun. The model shown here is the HK13E. Although usually fed from a box magazine, the HK13E can be converted to belt feed by fitting a magazine adaptor and a different bolt. The HK13E also features the 3-round burst mode of fire in addition to the usual single-shot and continuous modes.

Left: A unit of Portuguese troops, armed with an HK21 machine gun, move through a village in Guinea Bissau in search of guerrillas.

GALIL ASSAULT RIFLE

Small countries usually find it more economical to buy their weapons from major manufacturers, but Israel is a special case. Its geographical isolation in a sea of shifting loyalties means that it cannot rely upon uninterrupted supplies from any outside source, and it has therefore had to develop its own armaments industry. One of the Israelis' earliest priorities was to produce a reliable and accurate rifle as the basis of their armoury, and in the Galil they certainly achieved their aim. There is little in the world of firearms that is totally new, and the fact that some features of the Galil were adapted from other designs does not detract from its excellence.

When the Israeli Defence Force (IDF) came into being in 1948, it was armed with whatever it could obtain, and the infantry rifle was usually the British .303in Lee-Enfield or German 7.92mm Kar 98k. In the 1950s they were replaced by the well-known FN FAL rifle in 7.62mm Nato calibre, a choice which was economically sound and which ensured that irrespective of political alignments there would always be a source of spare parts and ammunition. But after the Six-Day War of 1967 it was decided to equip the Israeli Army with a modern 5.56mm rifle, and competitive trials of several designs were held. The American M16 and Stoner 63, the Soviet AK-47, a proposal by Uziel Gal (designer of the famous Uzi sub-machine gun), a design by Israel Galil and Yakov Lior, and one or two others were all put through their paces. The field trials were exceptionally severe and none of the prototypes met the stringent requirements laid down by the IDF, but the Galil was nearest and the inventors won the annual Israeli Defence Award in 1969.

The next two or three years were taken up in modifying and retesting the rifle until in the spring of 1973 it was formally adopted as the official service rifle of the IDF, replacing the FN FAL. In addition it replaced the FN general purpose machine gun in the squad automatic role, as well as replacing the Uzi sub-machine gun in some other roles. Production, however, was slow to get under way and the Galil was not in service in any great numbers in time for the Yom Kippur War of October 1973. It has, though, seen a good deal of combat employment since then.

The Galil is a gas-operated weapon, which relies heavily on the AK-47 and its Finnish variant, the Valmet M62, since the general system of operation is identical. There is a gas cylinder above the barrel in which a piston-rod operates. Between the barrel and the cylinder is a fixed-aperture gas port, through which gas is tapped from behind the bullet to drive the piston backwards. The piston bears on the end of the bolt carrier, the upper part of which is tubular and acts as a seat for the return spring. The lower part carries the bolt, and inside this carrier is a cam slot engaging with a pin in the bolt. As the bolt carrier is driven back, so the cam slot causes the pin to move sideways, rotating the bolt to unlock it from the breech. As soon as the bolt is unlocked, the continued rearward movement of the carrier withdraws it, extracting the spent case and ejecting it. The final movement of the bolt and carrier assembly forces the firing hammer downwards until it is retained by the sear. The return spring then expands, driving the carrier forward; the bolt collects a fresh cartridge from the magazine, mounted below the receiver, and feeds it into the chamber. The bolt comes to a stop, but the carrier continues so that the cam slot now turns the bolt in the opposite direction, locking its lugs into recesses in the barrel. Pressure on the trigger then releases the hammer, which rises to strike the firing pin and begin the sequence over again.

There is a selector switch on the side of the rifle which has three settings: safe, automatic and repetition. When set at automatic, the action is as described above, but the hammer is held by an auxiliary sear until the bolt has closed and locked, whereupon the auxiliary sear is released by the movement of the carrier and the hammer comes up to fire the next round; the firer has held the trigger pressed during all this.

Previous page: Two Israeli infantrymen, armed with Galil ARM rifles, stand in front of a UN observer in Beirut. Above: Israeli troops take their ease in Lebanon; the soldier kneeling is armed with the SAR version of the Galil. Left: An Israeli soldier prepares to loose a rifle grenade from his Galil. Note the small 12-round magazine. Right: Israeli troops rest, with ARMs alongside, bipods lowered.

Additional features of the Galil include a bipod (which can also be used as a wire-cutter), a flash hider on the muzzle, and a folding shoulder-stock of tubular steel. The foresight is a post, protected by a circular hood, and the rear sight is a two-position flip aperture set for 300m (330yds) and 500m (550yds) range. In addition there are night sights: two small leaves, one above the rear sight and one behind the foresight, which can be flipped up to expose Tritium luminous spots, two at the rear and one at the front. By lining up these three dots with the target, quite reasonable shooting can be done in poor light.

There are three magazines – based on the US Stoner magazine – for 12, 35 and 50 rounds. The 12-round magazine is intended only for use with grenade-launching cartridges, while the 50-round magazine is for use when the Galil is being employed as a light machine gun. In the normal rifle-role the 35-round magazine is standard.

This basic model of the Galil is known as the ARM (assault rifle and machine gun) model, and it was soon followed by two modified versions: the AR (assault rifle) which is the same as the ARM but without the bipod; and the SAR (short assault rifle) which is the same as the AR but has a barrel 33cm (13in) long instead of the 46cm (18in) of the AR. When the shoulder-stock is folded the SAR becomes a very convenient sub-machine gun.

In the late 1970s a 7.62mm model was developed, principally for export. This is precisely the same as the 5.56mm Galil except for the larger dimensions demanded by the larger bullet. The AR/ARM version has a 53cm (21in) barrel, and the SAR has a 43cm (17in) barrel, and both use a 25-round magazine.

In 1983 a new sniper model was announced. This uses the basic Galil action, in 7.62mm Nato calibre, but there are a number of features special to the model. The bipod is mounted behind the fore-end and attached to the receiver enabling it to be easily reached and adjusted by the firer and, in addition, ensuring that the barrel is not stressed in any way. The barrel is heavier than standard, contributing to the accuracy. The mounting for the telescope sight is on the side of the receiver and is a precision-cast long-base unit giving good support to the 6×40 Nimrod telescope sight supplied as standard. Both mount and sight can be quickly dismounted and replaced without disturbing the zeroing, and any type of night sight can be fitted to the mounting. The barrel is fitted with a muzzle brake and compensator which reduces jump.

The ARM with stock open (top) and – from the reverse side – with stock closed and carrying handle folded over (above centre). Above: The 5.56mm ARM Galil weapon system – three magazines of 12, 35 and 50 rounds capacity; bayonet and scabbard; three different types of rifle grenade. The carrying handle is in the up position and the selector switch is on 'S' for safety. Alongside the US M16 and the Belgian FN FNC the Galil ARM is a leader in the field of 5.56mm assault rifles.

Main picture: Armed with an array of Galil assault rifles, Israeli troops parade by their national flag. The Galil proved itself to be a highly effective combat weapon during the Israeli invasion of Lebanon in 1982.

Left: The simplified AR version of the Galil – without bipod and carrying handle – shown here with its stock folded.

The short-barrelled SAR Galil with stock open (left) and folded (below left). Below: The 7.62mm sniper variant of the Galil. Besides the powerful telescopic sight, other features of this weapon include a special sharpshooter's rifle sling, a folding wooden stock with recoil pad and cheek piece, and a two-stage trigger. This model employs the standard 7.62mm Nato round.

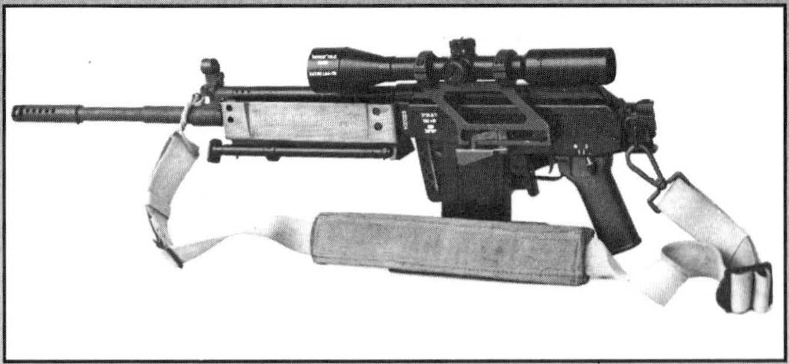

Right: Israeli troops armed with two very different 5.56mm smallarms – the Galil ARM and the M16. Below: With APCs at the rear, Israeli troops stand guard armed with Galils.

Galil Assault Rifle

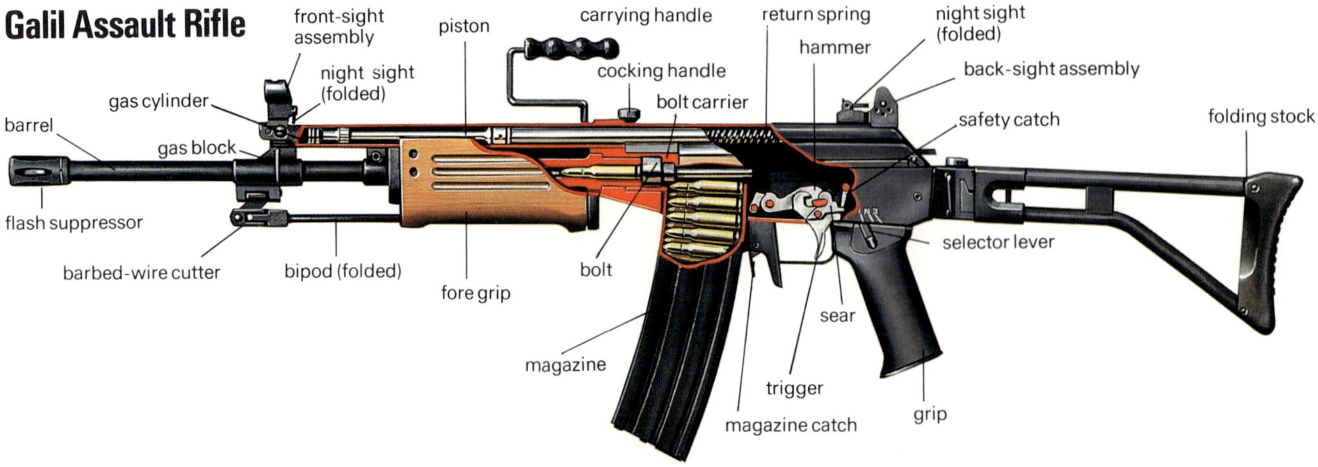

Galil AR/ARM Assault Rifle

Calibre 5.56mm
Length 97.9cm (38½in) with stock extended; 74.2cm (29in) with stock folded
Weight (loaded with 35-round magazine) 4.91kg (10.8lb)
Rate of fire Cyclic 650rpm; auto 105rpm; semi-auto 40rpm
Maximum effective range 500m (550 yds)
Ammunition Ball, blank, grenade-launching
Magazine 12-, 35- and 50-round box
Cartridge M193 5.56 × 45mm
Muzzle velocity 980m/s (3215fps)

Above: A group of South African infantrymen stand beside a mortar armed with R4 rifles, the locally-manufactured version of the Galil. Below: Although very similar to the Israeli Galil, the R4 has a number of different features, most notably a redesigned hand guard, a longer butt and the use of carbon-fibre.

It is possible to remove this unit and replace it with a silencer if desired. There is a two-stage trigger without provision for automatic fire. The stock is of wood, but it can still be folded forward in order to reduce the overall length for stowage or travelling. The butt is fitted with a recoil pad and cheek-piece, both of which can be adjusted to suit the firer. During trials, all shots fired at 300m (330yds) range came within a 15cm (6in) circle.

The Galil rifle has been purchased by a number of armies. One interesting adaptation has been by the South African Army who have lengthened the butt, to better suit the larger physique of South African troops, and have replaced the tubular steel with a carbon fibre material which is stronger than steel and which does not get so hot in the high temperatures found in the African bush. This is now in full production in South Africa as the R4 rifle. The Dutch company of NWM has shown a considerable interest in the Galil and has been granted a licence to manufacture the Galil as the MN1. As with the R4, the MN1 fires the 5.56mm round.

M16 ASSAULT RIFLE

The M16, or Armalite rifle, as it is still often called, is a weapon which can be found all over the world. After a bad start, largely due to poor instruction of the soldiers who had to use it, it has been improved into its present M16A2 version which began issue to the US Army in 1985. It was responsible for the universal acceptance of the 5.56mm cartridge which has now become Nato standard, and its magazine design, although far from perfect, has been taken as the model for newer rifles and machine guns. Its appearance made plastic respectable in rifle design but it is still a controversial weapon. Those who love it acclaim it as 'the world's greatest combat rifle'; those who hate it, on the other hand, refer to it as 'the Baseball Bat'.

In the early 1950s, in the aftermath of the Korean War, the United States Infantry Board decided that a new, lightweight rifle was required, and in 1957 they published a requirement which asked for a weapon weighing not more than 2.72kg (6lb) when loaded, as accurate as the Garand M1 rifle at 460m (500yds) range, capable of automatic or single shot fire, and capable of penetrating a steel helmet or body armour at 460m (500yds).

Among the companies who took notice of this was the smallarms research company Armalite Incorporated of Costa Mesa, California. Their chief designer, Eugene Stoner, had already helped develop a modern gas-operated automatic rifle known as the AR-10, firing the 7.62mm Nato cartridge, but the weight requirement of the new weapon meant that a small high-velocity cartridge had to be used. The US Army and Air Force had been involved in an experimental programme which had led them to believe that the existing .222 Remington cartridge held promise, and Stoner began working on what was virtually a scaled-down AR-10 in .222 calibre.

Trials with the first prototype showed that the .222 cartridge was causing excessive chamber pressure, so the round was redesigned to have a slightly larger case, thus providing space to make adjustments to the propelling charge. This became the '.223 Remington' and has since been adopted world-wide as the 5.56×45mm cartridge. In March 1958, 10 test rifles were delivered to the US Army, and after extensive trials the new rifle, now known as the AR-15, was recommended as a possible replacement for the existing M14, though, generally, throughout the army there was strong opposition to the whole concept of small-calibre weapons. Subsequent tests led to some minor changes in the design, and in 1959 Armalite came to a business agreement with the major arms manufacturer Colt for the latter to manufacture the AR-15 in quantity.

The first large order for the AR-15 came from the

Previous page: An M16-armed Gurkha in Brunei. Above: The M16 in action with Marines in Vietnam. Below: An M16 with Starlight night scope and XM148 grenade launcher. Opposite top: British and Gurkha troops in Hong Kong. Far right: An AR15 fitted with the Sopelem OB50 night sight. Right: South Vietnamese forces bring in a prisoner.

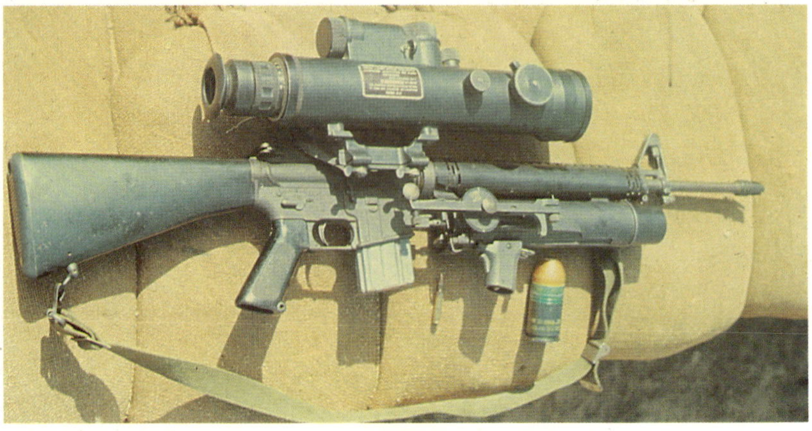

US Air Force, which required a light weapon for guard patrols on strategic bombers and airbases, and in 1962 8500 rifles were ordered. Then the Department of Defense ordered 1000; in 1963 a further 104,000 rifles were ordered for use by the air force and other categories of special service troops.

Many of the US Air Force weapons went to Vietnam, and these, together with the 1000 which had been sent to the US Military Assistance Group there, found immediate favour with the South Vietnamese who, being rather smaller than the average American soldier, found the light AR-15 a more convenient weapon. The Vietnamese requests for more AR-15s, combined with Defense Secretary Robert McNamara's enthusiasm for the rifle, led to the army taking rather more interest; they realised it would be a good weapon for the type of close-combat typical of the fighting in Vietnam, and by 1966 the army had bought more than 400,000 rifles. Shortly afterwards it was standardised as the M16 rifle.

The M16 has, for a rifle, very little recoil, due to an advanced method of operation. It is gas operated, without the use of a conventional piston. When fired, a portion of the gas following the bullet is tapped off through a hole in the barrel and directed down a stainless steel tube back into the receiver where it is expelled into the hollow bolt carrier. The expanding gas forces the carrier back, and a cam slot in the carrier, working on a lug on the bolt, causes the bolt to be revolved to unlock from the chamber, and then be pulled back together with the carrier. As the bold unit goes back so it re-cocks the hammer and compresses the return spring. At the end of its recoil stroke, the spring exerts itself and drives the bolt forward to collect another round from the magazine and drive it into the chamber. The bolt head comes to a stop on the cartridge but the bolt carrier continues forward for a short distance so that the cam slot turns the lug and locks the bolt. The firer then presses the trigger, the hammer is released to strike the firing

M16 Assault Rifle

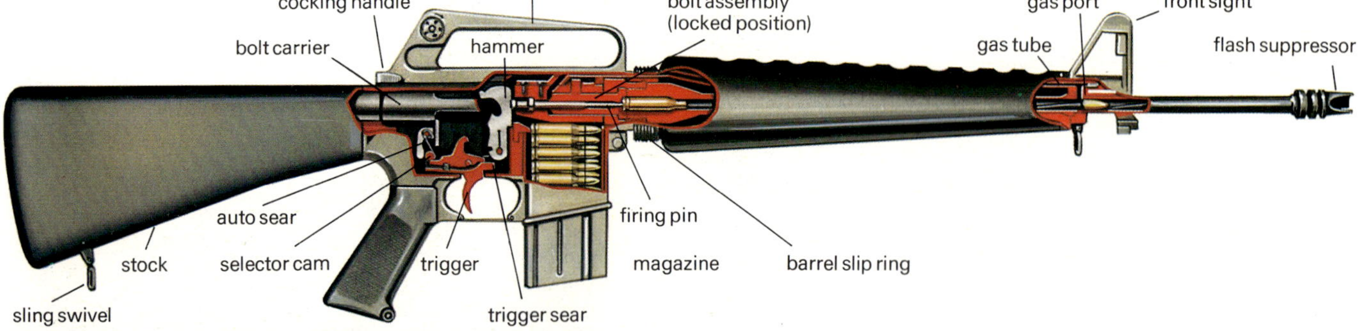

M16 Assault Rifle

Calibre 5.56mm
Length 99cm (39in)
Weight (loaded with 30-round magazine) 3.82kg (8.42lb)
Rate of fire Cyclical 700-900rpm; practical automatic 150-200rpm; practical semi-automatic 45-65rpm
Maximum effective range 460m (500yds)
Magazine 20 or 30-round box
Cartridge M193 5.56×45mm round
Muzzle velocity 1000mps (3280 fps)

Left: Armed with an M16A1, one of Charlie Lademora's troops patrols the jungle area of the Tambis region of south Mindanao in the Philippines. Lademora, a former officer with the Philippine Army, set up a fighting force of army deserters and criminals in the mid-1970s to combat the threat of communist infiltration in the area. The M16 shown here mounts the standard US Army M203 grenade launcher, considerably increasing the firepower available to the unit, and is fitted with the later, curved, 30-round magazine.

Above: Equipped with M16A1s and the telescopic-stock variant, the M16A1 carbine, a unit of the 1st Air Cavalry Division (Airmobile) is deployed from a UH-1 against Viet Cong positions.

Below: US airborne troops capture a hilltop in Vietnam. Their M16s were a simple and effective weapon, particularly suited to the close-fire fighting typical of the war.

pin, and the next shot is fired. On automatic fire, of course, the hammer rises as soon as the bolt is locked and fires the next round without the firer having to release and re-press the trigger.

When the M16 was introduced, much was made of the simplicity of the design and the fact that it required minimal maintenance and cleaning. Unfortunately, the troops in Vietnam took this to mean that it needed no cleaning at all, and this led to a spate of complaints about failure to eject and various other types of malfunction. A Congressional enquiry found that the type of powder used in the original 5.56mm cartridge had been changed to that used in the M14, a variety which generated much more fouling and carbon, and when deposited in the bolt mechanism it solidified to a rock-like hardness which could jam the weapon. Some small changes were made to the rifle, notably the chroming of the chamber and later the whole barrel, thereby preventing the build-up of rust which had previously been a serious problem. A plunger was fitted, allowing the soldier to drive the bolt home if the cartridge was reluctant to enter a dirty chamber and the necessary cleaning equipment was issued. Alongside these improvements an education programme pointed out the need for basic standards of weapon care. After that the rifle was known as the M16A1 and thereafter worked effectively in the field.

During the 1960s much ill-informed comment was published about the lethal effect of the 5.56mm bullet fired from the M16 rifle. It is a light bullet which, in order to have the maximum possible target effect, has been closely matched to the rifling which spins it. So

Above: One of Lademora's cowboys cradles a Philippine version of the Colt carbine. Like the M16A1, the carbine is provided with a forward bolt assist directly below the cocking handle. The weapon shown here is fitted with the standard M16A1 closed flash suppressor and a 30-round magazine.

Below: The XM177E2. This later version of the Colt Commando was fitted with a 292mm (11.5in) barrel to permit grenade launching but retains the retractable butt and the 101mm (4in) flash suppressor of the E1.

long as the bullet is passing through the air it is stable and accurate, but as soon as it meets something thicker than air it will tumble over and deliver up all its energy in a massive blow. Thus, during the early stages of the Vietnam War, rumour circulated through the army about the explosive impact of the new small-calibre bullet. But while the M16 bullet was capable of inflicting severe damage to the human body the reports of Viet Cong being killed by shock from wounds in the hand were undoubtedly exaggerated. In addition, the latest 5.56mm bullets are somewhat heavier than the original American design and require a faster spin, so that they are now more stable and less devastating in their effect on the body.

During the Vietnam War a shorter and handier version of the M16 was issued to special forces units in the American armed forces. Initially designated the XM177E1 (XM177 for the air force) it became better known as the Colt Commando or CAR-15, and was highly regarded by the combat soldier in Vietnam. Although the firing mechanism is the same as that of the M16, instead of the standard 533mm (21in) barrel the CAR-15 has a cut-down barrel only 254mm (10in) in length, and is provided with a telescopic butt which can be pulled out for firing from the shoulder.

Lacking the accuracy of the M16 at longer ranges, the CAR-15 was in effect a small-calibre submachine gun and as such proved ideal for the close-range jungle fighting that was characteristic of the war in Vietnam. Besides reducing long-range accuracy, the CAR-15's shorter barrel also had the effect of dramatically increasing muzzle flash so that a 101mm (4in) suppressor had to be attached to the end of the barrel, though a later model, the XM177E2, was fitted with a longer barrel which reduced the need for a flash hider. Although manufacture of the CAR-15 ceased in 1970 it remains a highly potent weapon.

In order to provide the infantry section with more punch, Colt developed the 40mm XM148 grenade-launcher which was designed to be attached to the underside of the M16 barrel. While popular with the GIs, the launcher was not considered sufficiently reliable and was replaced in the 1970s with the M203, also of 40mm calibre. Capable of delivering a rifle grenade to a distance of up to 500m (460yds) the M203 remains as the standard US Army launcher.

Since its standardisation with the US Army the M16 has been widely adopted in other countries; it is currently in use by the armies of Chile, the Dominican Republic, Haiti, Italy, Jordan, South Korea, Mexico, Nicaragua, Panama, the Philippines and Vietnam. More than 4 million M16 rifles have been manufactured in the USA and – under licence from Colt – Singapore, the Philippines and South Korea also produce the rifle. Numbers are also owned by other armies, including the British, who issue them to special forces, and used them in small numbers during the fighting in Borneo in the 1960s.

HANDGUNS

The position of the handgun in infantry armament is something of a puzzle; officially it is the armament of officers and specialists, but experience shows that people on the battlefield with pistols are often paid special attention by snipers, so most officers prefer to carry rifles. Pistols are often carried by mortar men, missile operators and drivers as a last resort for self-defence, although the sub-machine gun is often preferred on account of its superior firepower. In spite of the fact that many armies do not issue pistols down to the lowest ranks, it is acknowledged that many soldiers will somehow manage to acquire a pistol during the course of a war.

Modern handguns may be divided into two main categories, the revolver and the automatic, each of which has its own advantages and disadvantages. The central feature of the revolver mechanism is the cylinder – chambered to hold on average six cartridges – which revolves to bring a fresh chamber to the firing position with each operation of the trigger. There are relatively few moving parts, and the revolver is therefore reliable and easy to maintain. In comparison with the automatic, however, it is bulky, which under some circumstances may be a drawback.

The compactness of the automatic, or more accurately the semi-automatic, derives from its method of operation, which has, however, the disadvantage of being more complicated than that of the revolver. Ammunition is held in a magazine, usually in the butt of the pistol, and fed by pressure from a simple spring mechanism to the firing position with each action of the slide. The slide normally operates on the blowback principle, being propelled backwards along the barrel of the pistol, which it houses, by the recoil force of the discharged cartridge. Although there are some automatics which operate by different methods, most feature a variation on this blowback principle.

Since World War II, the tendency has been for the military to adopt the automatic as the standard-issue side-arm, while the greater reliability of the revolver has recommended itself to police forces.

In military service, a pistol is of little use on the modern battlefield where the longer range of assault rifles easily exceeds the 40-50m (45-55yds) effective range of a trained marksman armed with a handgun. It remains useful only for those combat personnel operating in a confined space such as aircrewmen or tank crewmen simply because there is only enough room for a small weapon. A handgun may also be useful in close-quarter combat such as fighting in built-up areas or in trenches, but only when nothing better is available. This being so, the fact that effective handguns were developed in the first 30 years of this century has meant that development since then has been slight.

Of the automatics now in military service, the Colt ·45 M1911/M1911A1 is the most venerable design, dating from the first decade of this century. Still the standard-issue side-arm of the United States armed forces, the M1911A1 is almost the only military pistol to retain the ·45 cartridge, and it is universally regarded as well overdue for replacement, many of the 418,000 still in service having been virtually rebuilt many times since their manufacture. To date there has been at least one unproductive competition in order to find a replacement for the M1911A1, but the results have been inconclusive, many would say for political reasons. It is still planned to replace it with a 9mm Parabellum automatic, to be known as the XM9, but the Colt ·45 seems certain to stay in use for some time.

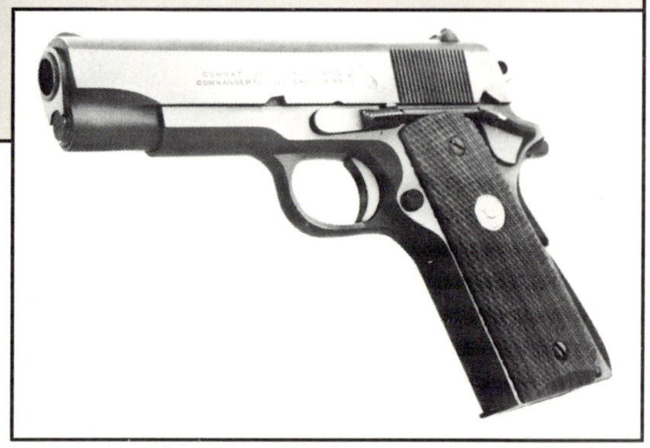

The Browning-designed Colt M1911 ·45 automatic pistol (above), produced in huge quantities since its adoption by the US in 1911, has been the basis for many variants, including the Combat Commander (right).

Previous page: A Sandinista guerrilla in Nicaragua, armed with a Smith & Wesson automatic. Though of limited military value to conventional armies, the handgun is an important weapon in the arsenal of many insurgent movements. Often more easily obtainable than heavier weapons, it has the advantage of being easier to conceal for clandestine operations.

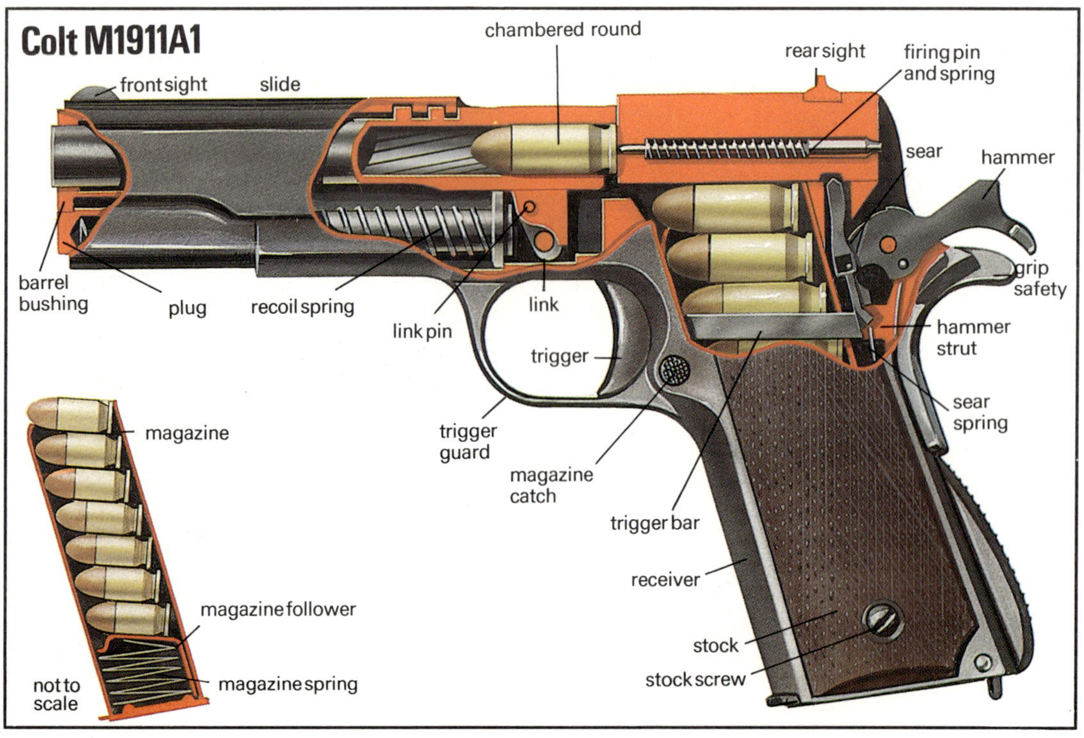

Right: A classic design, the Walther PP, introduced in 1929, has been a major influence on subsequent handgun development on both sides of the Iron Curtain. Far right: The Walther P5, like the PP designed specifically for police work, is a cut-down version of the P1, itself a modern development of the wartime P38.

Above: The 9mm Browning High-Power, standard pistol of the British and many other armies. Right: The Walther P38, World War II handgun of the Wehrmacht. Below: The P38 in use in the 1970s with Rhodesian reservists.

The Colt M1911A1 is not the only veteran design still in production. The Browning 9mm High-Power, patented in 1927, and like the M1911A1 designed by John M. Browning, is produced around the world in many versions, and must outnumber virtually all other pistols in use today. Despite its weight and bulk (the magazine holds 13 rounds), it is still one of the best service pistol designs. It is strong, reliable, and capable of absorbing hard knocks without malfunctioning. Production rights to the Browning design are held by FN of Belgium, who produce several 'special' versions, including a cut-down model.

Another group of pistols whose design originates from the pre-World War II period are those produced by Walther in Germany. The Walther PP was originally designed as a police weapon (Polizei Pistole – hence the initials), and was introduced in 1929 while the PPK (Polizei Pistole Kriminal) was introduced in 1931 for use by plainclothes policemen, being a smaller version of the PP and hence more easily concealed. Both pistols fire the 7·65mm round and are excellent weapons which are still widely used by many police and military forces, including the British Army.

Walther also produce the P1, a modern version of the World War II 9mm P38. The P1 is lighter than the P38, which first introduced many of the features now common in modern automatic pistol designs. The P1 has an external hammer, which may be operated by either single or double-action, several safeties, and a button next to the rear sight, which rises when the chamber is loaded. The P1 now uses light alloys in its construction, a common feature with many modern pistols, which along with the use of plastic components can considerably reduce the weapon's weight. Walther also produce the 9mm P5, which was designed around a West German police specification, and may be regarded as the modern equivalent of the PP. It embodies several features from earlier designs, allied to some new safeties which make the P5 a very easy and safe pistol to use.

Apart from the Browning and Walther designs, many other pistols in use in the 1980s incorporate principles which date back several decades. A typical example is the Swiss SIG range, which is widely regarded as being the finest made today. The basic design principles of the SIG pistols, however, come from a French Petter design, first produced well before World War II. Over the years these pistols have been developed to a remarkable degree, and many

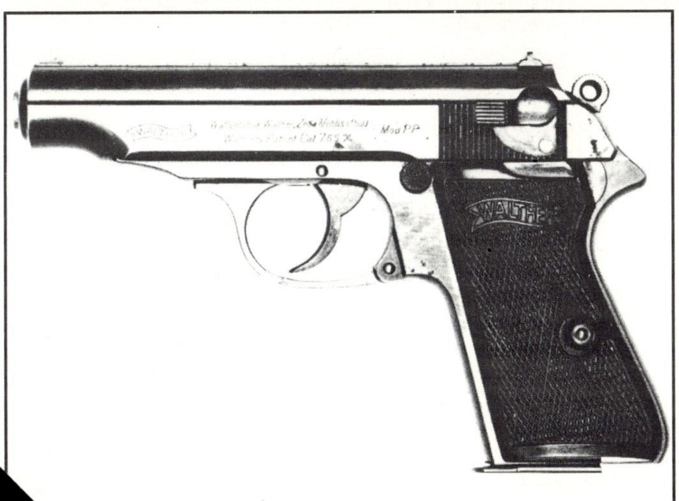

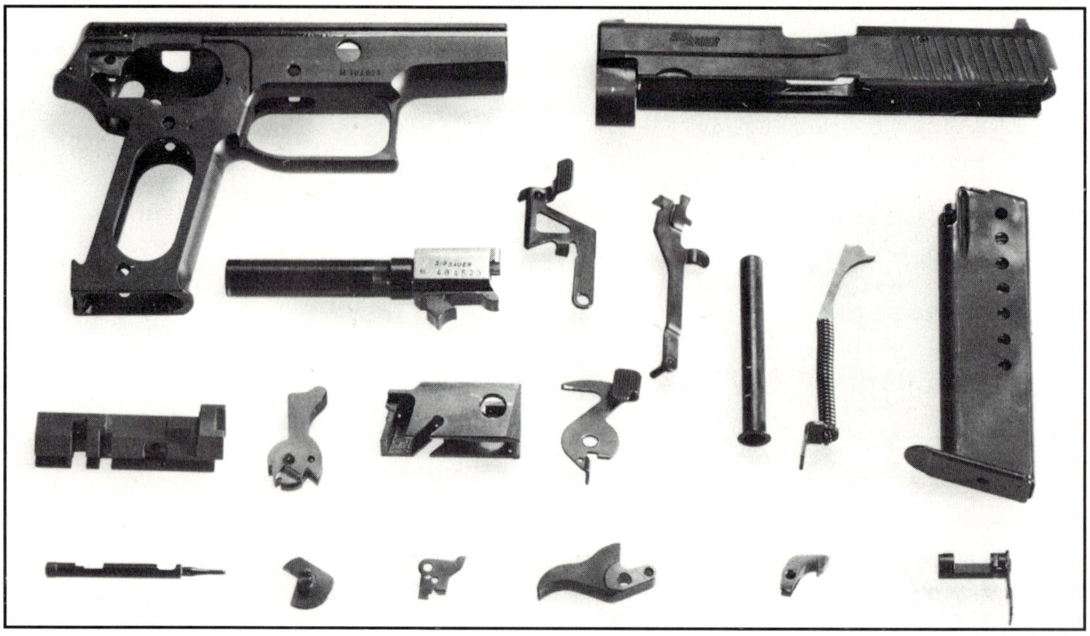

Right: A stripped-down 9mm SIG-Sauer P225 pistol. The complexity of its operation, and the relative difficulty of maintenance, are a drawback of the automatic when compared to the simple, easily maintained revolver.

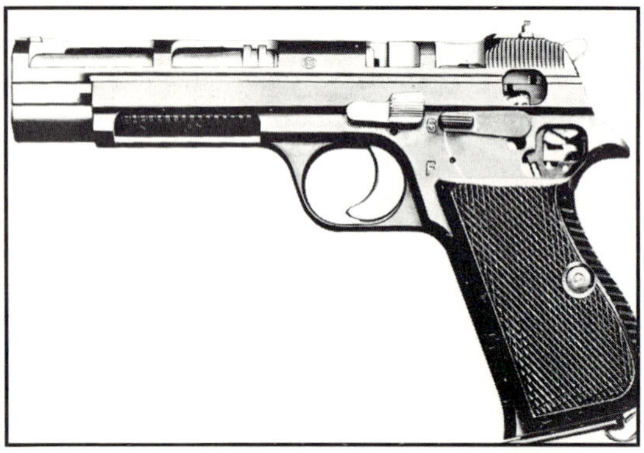

Left: The Swiss SIG 9mm P210, part of a fine family of pistols which operate on the French Petter principle.

Left: The German firm of Heckler & Koch produce a range of pistols, including this P9S, which share a characteristically modern appearance, and are rugged and easy to maintain.

refinements have been added. In recent years, SIG have joined with the West German Sauer concern to produce even better pistols.

One of these is the P226, which was one of the original entries in the US armed forces' XM9 contest. The P226 is now one of the most sought-after of all modern pistols for its balance, finish, and overall practicality. One of the earlier SIG models, the P220, is in service with the Swiss Army, as well as being produced for export, as is the P210. All of these pistols feature an external hammer, which can be cocked by the thumb if required. If the pistol is cocked, but not fired, it is possible to apply a safety mechanism which allows the hammer to be lowered onto the firing pin without discharging the pistol. This is now a common requirement for modern automatic pistols, which was lacking in many earlier designs.

Another German manufacturer is Heckler & Koch, whose pistols range from the HK4, through the P9S and the VP70M (which might possibly be regarded as a form of machine-pistol), to the advanced P7. The common feature of all Heckler & Koch pistols is their striking appearance, and all designs make use of the same curved, sweeping lines. This actually contributes to their utility, as the smooth outlines prevent snagging on clothing. The Heckler & Koch pistols embody all the demands of modern armed forces, including all the usual safeties, ease of maintenance and cleaning, few moving parts, and robust construc-

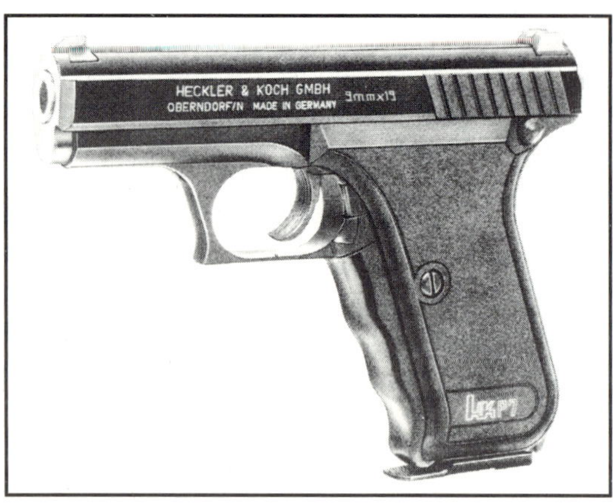

Left: The Heckler & Koch P7, in service with the West German Bundeswehr, is a tough 9mm combat pistol.
Right: The much-praised Steyr GB incorporates a chrome barrel with hexagonal rifling.

tion. Typical of Heckler & Koch designs is the recent P7, which was originally designed for use as a police weapon, but has now also been adopted by the West German Army (Bundeswehr). It is a very simple and rugged weapon, with few moving parts to jam or break, and it can be stripped down for cleaning in seconds. Like many modern pistols, it employs the 9mm x 19 Parabellum cartridge.

In Austria, Steyr-Daimler-Puch have for many years been producing superlative weapons of all kinds, and their GB 9mm pistol is no exception. It is a superb pistol, although it has yet to find a large-scale market. It is unusual in many ways, the most important being that exhaust gases, bled off from the barrel after firing, are used to delay the rearward progress of the slide, in order to ensure safe locking. This principle was first used in the World War II German Volksgewehr, and though rarely used since, is a safe

Right: The most common handgun with Warsaw Pact forces, the Soviet 9mm Makarov derives many of its features from Walther designs.

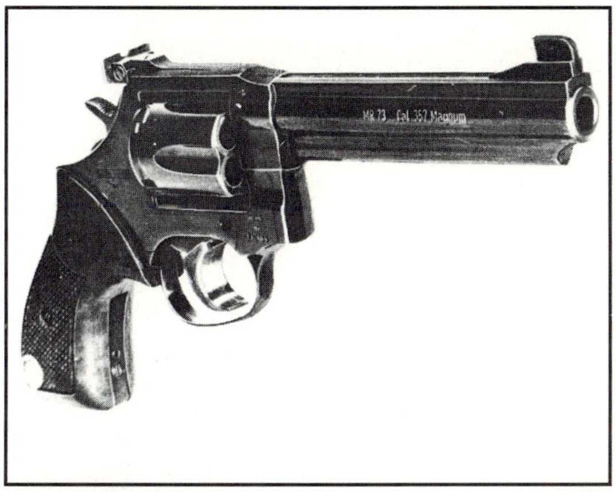

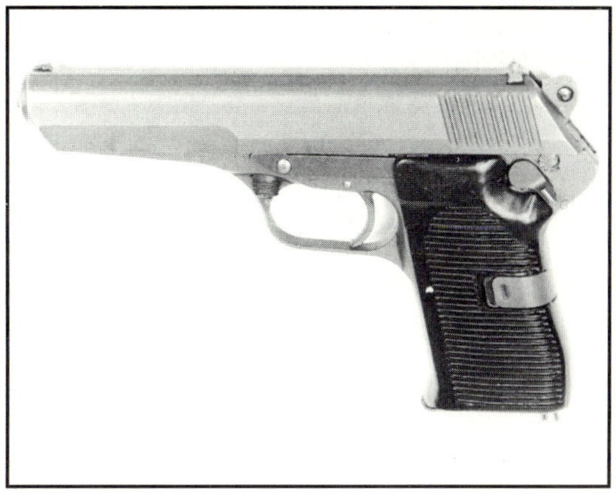

Left: Typical of many modern combat revolver designs is this French Manhurin ·357 Magnum MR73. Right: The Czech 7.62mm CZ 52 was adopted by the Czech Army as an alternative to the Makarov. The Czechs have managed to maintain their own arms industry, despite the almost universal employment of Soviet weapons in the Warsaw Pact.

Left: The Manhurin is in use with the Groupe d'Intervention de la Gendarmerie National (GIGN), the French anti-terrorist unit. Right: Another Warsaw Pact variation on the Walther theme is this Polish P64 9mm pistol, in service with the Polish Army.

Right: The French MAB PA15 is another pistol which employs the Petter principle. It is the standard handgun of the French armed forces.

and simple locking method. The barrel of the GB is chromed, in order to reduce wear and to ease cleaning, and in place of the normal rifling, a smooth form of hexagonal twisting of the bullet is used. The magazine in the pistol grip can hold 18 rounds, which is useful in combat, yet the GB sits easily in the hand, and can be fired with none of the 'jumpy' recoil normally present in a 9mm pistol.

In Eastern Europe, the Soviet Union has long been unadventurous in pistol design, and currently relies upon the 9mm Makarov for its standard service side-arm. The Makarov is unexciting in overall design and handling, but in common with all Soviet smallarms, it is very sturdy and well able to take the trials of Red Army service life.

In recent years, however, a new Soviet pistol has emerged, in the form of the 5·45mm PSM, which is a small weapon, perhaps intended for use by the security services. The cartridge is very low-powered, however – it would appear to have even less power than a normal ·22 round. The pistol resembles the Walther PP in appearance, and is not yet in service in large numbers.

Not all Warsaw Pact forces use the Makarov. The Czechs favour their own Model 52, calibered in 7·62mm. This is a sound design, though unspectacular, and was adopted by the Egyptian Army as well as by the Czechoslovak Army, but it is no longer in production. Czechoslovakia now produces the 9mm Model 75, but this has yet to find a customer.

The Polish Army employs its own 9mm P64, which may be regarded as an up-dated Walther PP. The P64 is an attractive and compact design, with few frills, and may be fired using single- or double- action.

In France, the Manufacture d'Armes de Bayonne (MAB) continues to produce several pistols based on the Petter principle, which dates back to well before World War II. Nevertheless, all the MAB pistols are strong and serviceable weapons, one of which, the PA15, is now the standard pistol of the French armed forces. Manhurin, another French company, produces several types of revolver, including the ·357 Magnum MR73, which is in many ways typical of such weapons being produced today. It has a six-round cylinder, which swings out sideways for reloading, and it features a single- or double-action trigger mechanism.

In Italy, Beretta maintain their traditional high standard of automatic pistol designs. During World War II, a Beretta pistol was a highly-prized trophy, and modern Berettas are no less sought after. It was a Beretta design, the Model 92, which was deemed to be the best entrant in the abortive XM9 contest, but it is only one in a whole range of excellent weapons. Most of the modern Beretta pistols use either a 13- or a 15-round magazine, and differ in the types of safety fitted, as well as in other small details. Some are chambered to take the 9mm Short cartridge, but the Model 92 and its variants are chambered for the ubiquitous 9mm Parabellum.

Meanwhile, what of the world's largest pistol-producing nation: the United States? The two giants of the industry, Colt and Smith & Wesson, continue to produce an array of revolver designs, many of which seem to differ only in the markings applied. Many small companies find it profitable to manufacture spares for the ever-popular M1911, or to devise all manner of imaginative 'commando' variants. The Ruger company is making considerable inroads into the revolver market with a range which covers several calibres, nearly all of which are manufactured from stainless steel. Ruger pistols are now in service with police forces and paramilitary formations throughout the world, but have few innovatory features, apart from their seemingly indestructible construction.

Only Smith & Wesson offer any novel designs, including a pair of 9mm automatics, known as the Models 459 and 469. These can trace their ancestory back to the Model 39 of 1954, but recently, no doubt spurred on by the impetus of the US Army's XM9 programme, the design has been refined considerably. The Model 459 is now a 14-shot pistol, with an aluminium frame, while the Model 469 is a slightly shorter version, holding only 12 rounds.

Below: The classic Beretta Model 1934 7·55mm automatic. The wartime service pistol of the Italian Army, it was a sought-after souvenir for Allied troops. Below left: A modern design, the Beretta Model 92 is an excellent combat pistol. Below right: The Model 92's 15-round magazine. Bottom: One of the few modern American designs, the Smith & Wesson Model 459, is a 14-shot aluminium-framed pistol.

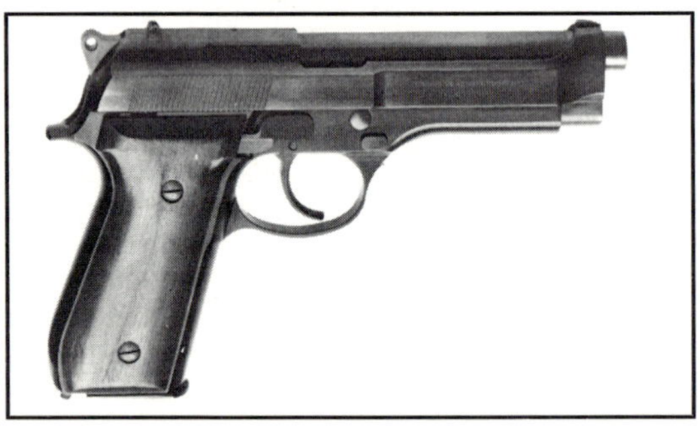

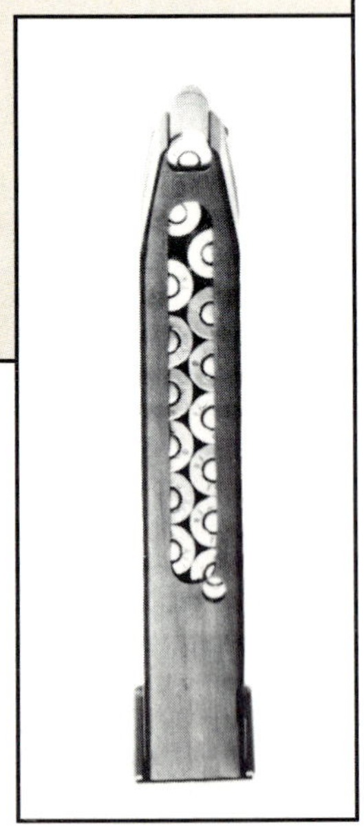

APPENDIX A
ELECTRONIC WARFARE SYSTEMS

Throughout the history of land warfare, the commander with the better lines of communications has always had a major advantage over his opponent; however sophisticated the weaponry deployed, it is severely hampered if the operator cannot be instructed when and where to fire, and if the commander cannot coordinate the actions of the various elements at his disposal. The emergence of reliable radio equipment during World War II revolutionised battlefield management; the success of the German tactic of Blitzkrieg depended heavily on the availability of effective electronic communications. Since that time, all the world's armies have come to rely on radio as their primary communication tool. Apart from communications equipment, another factor in the growth of the so-called 'electronic battlefield' has been the use of radar to detect personnel and vehicles on the ground and the use of direction-finding equipment to locate emitters of radio and radar signals.

There can be little doubt as to the effectiveness of available electronic equipment. This effectiveness has generated a need for counter-measures, to which radio and radar are extremely vulnerable. A communications net can be disrupted by jamming, and simply monitoring the transmissions of an enemy whose radio security is lax can reveal valuable intelligence. Also, a group of monitoring stations can be used to locate accurately a divisional headquarters, for example, by its radio transmissions, allowing it to be attacked and thereby paralysing an important subsection of the overall command structure. Both Nato and the Warsaw Pact have recognised the opportunities of electronic warfare and have provided their ground forces with an array of equipment designed to attack the enemy's sensory and communications systems and to provide signals intelligence (SIGINT).

The Soviet Army has adopted the name 'radio-electronic combat' for electronic warfare. They have organised specialised units at army and front level to carry out the electronic offensive. The basic units are the Signals Interception Battalion and the Radio-Electronic Combat Battalion.

The Signals Interception Battalion acts as an intelligence-gathering and radio direction-finding force. According to recent estimates, such a unit comprises 433 officers and men and 114 trucks, divided between an HQ Company and Radio Intercept, Radio Direction-Finding and Radar Intercept Companies. The Radio Intercept Company has an establishment of 97 officers and men and is equipped with 28 SR-50-M intercept receivers covering the VHF/UHF bands and aimed at FM-voice communications. The Radio Direction-Finding Company is slightly larger with an establishment of 118 officers and men operating 16 radio direction-finder sets drawn from the SR-19-V, SR-20-V and SR-25-V types. Together, these three sets can give accurate bearings on HF to VHF AM/FM-voice communications. The Radar Intercept Company has the largest establishment of the three, with 123 officers and men operating 15 radar intercept systems. These will be either the SB-20-V covering the 50MHz-11GHz frequency bands, or the SM-21-V covering the 50MHz-10GHz frequencies.

Above: The heart of the 'electronic battlefield' will be command, control and communications centres (C^3) – like this French installation – analysing information from the frontline.

The Radioelectronic Combat Battalion represents the offensive capability within the Soviet Army's electronic warfare units. These handle the actual jamming of enemy systems and, manned by 492 officers and men, are made up of three radio jamming and one radar jamming companies. The radio jamming units have an establishment of 105 officers and men operating four radio intercept sets, four direction-finders, 12 radio jammers and three multi-channel jammers. The radar company is manned by 93 officers and men and is equipped with 10 radar intercept/direction-finding sets and 11 jammers.

Both types of battalion are highly mobile, with all their equipment being either vehicle-mounted or towed. The threat posed by this capability, backed up

Left: A French infantryman with a TRC 743 field data terminal attached to his radio.

by further capacity mounted in Mi-4 Hound and Mi-8 Hip helicopters, is taken very seriously by Nato. Some observers believe that the Soviet system could reduce battlefield communications to the level of 1916, an extremely serious proposition when it is considered how reliant Nato is on radio communications. The capabilities of Soviet radioelectronic units were illustrated in Europe during the intervention in Czechoslovakia in 1968 when the Warsaw Pact forces involved employed a chaff and barrage-jamming techniques, forming a vast electronic screen along the borders between Czechoslovakia and West Germany. This effectively masked the troop movements and airlifts.

The main contingent of Nato's land electronic warfare units is provided by the US Army. In each of its 16 frontline divisions there is an organic electronic warfare/intelligence battalion which combines the roles of the Soviet units already described. The US Army has not yet adopted a standard table of organisation and equipment for its electronic warfare units. As might be expected, these battalions are furnished with a wide range of sophisticated equipment. Three important current systems are the AN/MLQ-34 TAC-JAM communications jammer, the AN/MSQ-103 signals intercept/direction-finding system and the EXJAM expendable jammer.

The first of these, the MLQ-34, is designed to

Left: A TEAMPACK AN/MSQ-103A installed on the back of an XM-1015. TEAMPACK is an extremely versatile system, which can be mounted on many different kinds of vehicles.

Right: The battlefields of the future will see great use of remotely-piloted vehicles, like this Compass Cope-R for reconnaissance and target observation as well as jamming and direction-finding. An RPV can carry a wide range of communications and surveillance equipment, tailored for specific mission requirements.

disrupt and deceive enemy communications links. The system as a whole comprises a number of receivers for target identification, high-powered transmitters and a controlling computer, the entire package being mounted on an M-548 tracked cargo-carrier chassis. Each transmitter is capable of generating a multi-kilowatt signal over a broad frequency range, making the system a powerful and flexible electronic warfare tool. Produced by GTE Sylvania, 17 TAC-JAMs have been completed and the US Army has a further 31 systems on order.

The MSQ-103 TEAMPACK equipment is a mobile system designed to identify and locate hostile communications transmitters. Used at the divisional level, TEAMPACK covers the 500MHz-50GHz frequency band and can be mounted on a range of vehicles including the XM-1015 tracked chassis, the M-35 utility truck and the jeep. Included within the system is built-in computer-processing and a wide-band data link for the transmission of received

Below: The launching of an Aquila remotely-piloted vehicle, the choice of the US Army for battlefield reconnaissance and electronic warfare. Made by Lockheed, it is still in the prototype stage, and had great difficulty meeting the US Army's requirements.

material back to a control/analysis centre. MSQ-103 is in current use with the US Army and will become the service's basic electronic warfare receiving equipment.

The EXJAM system currently under development by Fairchild Weston represents the leading-edge of American electronic warfare technology. Essentially a 'throw-away' system, EXJAM is made up of a number of transmitters mounted in a 155mm artillery shell. Once fired, each transmitter package is ejected at pre-set intervals and falls to earth along the flight path. On hitting the ground, the unit imbeds itself, deploys an antenna and begins transmitting a jamming signal almost immediately. EXJAM was field-tested during 1983/1984 and a manually-placed variant is also being developed.

Alongside its ground-based units, the US Army makes extensive use of airborne electronic warfare platforms such as the RU-21 series Beechcraft Utes, the EH-60A Black Hawk helicopter and the RV-1 series Grumman Mohawk. Such aircraft carry a wide range of systems including the Cefly Lancer and Guardrail signals intercept/direction/finding equipment on the RU-21, the Quick Fix jamming system on the EH-60A and the Quick Look surveillance system on the RV Mohawk.

The British Army operates a combined intelligence/electronic warfare system, believed to be concentrated in the 30th Royal Signals Regiment. This formation has the additional role of maintaining communications security, monitoring the army's own signals traffic to ensure that no information of value to an enemy is transmitted because of lax operator techniques. The British are reticent about discussing current operational equipment. Two

Below: Men of the US 82nd Airborne Division deploy communications equipment on Grenada in 1983. Bottom: The US TEAMPACK, here mounted on a Dragon 300 vehicle, is used to pinpoint hostile transmitters.

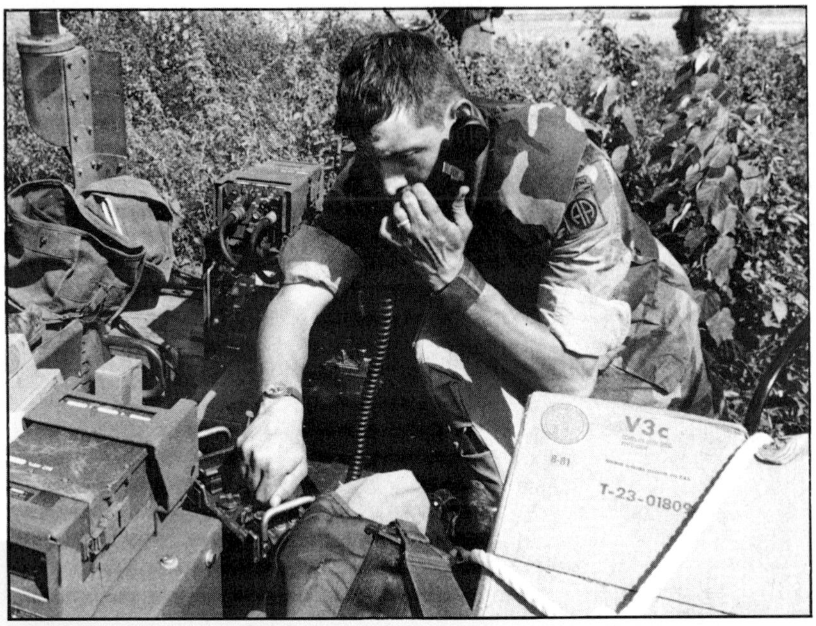

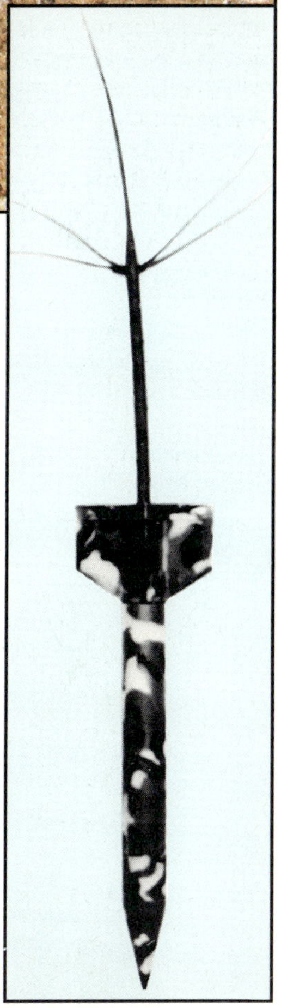

known examples of UK systems are the Marconi S373 and the MEL/Ferranti Barbican units. Both systems are vehicle-mounted, the S373 providing multi-band jamming and surveillance capability and the Barbican carrying out radar classification, analysis and direction-finding.

The S373 uses a number of high-powered transmitters feeding twin parabolic antennae mounted in a 'double bubble' radome which can be hydraulically raised or lowered. The system operators are housed in a container mounted behind the radome on the carrier vehicle, a standard 4-tonne Bedford truck. The system is also provided with a self-contained intercept capability and both voice and data-link communications channels.

Barbican uses a similar vehicle as a base and is highly automated, needing only a single operator. A full Barbican system would comprise three intercept stations, all joined by voice/digital data-link channels to a central control. Frequency coverage is believed to be between one and 18GHz. Both these systems are currently in the development stage and no information as to their eventual use in service has been released.

Outside Nato and the Warsaw Pact, the most experienced nation in electronic warfare is Israel. Precise details of Israeli equipment remain obscure, but it is believed that the country has a wide range of locally produced intercept and jamming equipment. These are highly mobile, being mounted on a variety of wheeled and tracked vehicles. Some examples of equipment are the EL/K7010 ground-based mobile intercept and identify radar emitters. Israel is unique in being the only country currently making operational use of an electronic warfare drone, the Mastiff remotely-piloted vehicle, which is employed for communications surveillance.

The future line of development of electronic warfare on land may be indicated by a new US project developing a digital electronic warfare capability for tanks. A threat-warning system incorporating sensors and display might provide greater survivability on the battlefield. The advantages of attaining 'electronic supremacy' over the opposition on the battlefield leaves little doubt that further developments in the use of electronic support and counter-measures will be forthcoming. Finding the enemy quickly will be especially important with the spread of precision-guided munitions, making a first strike more destructive.

Top: The Barbican intercept receiver, manufactured by MEL, is installed on a Bedford 4-tonne truck.
Above: An aerially delivered seismic intrusion detector (ADSID), used extensively by the US during the Vietnam War, especially to give early warning of an infantry threat to firebases.
Left: A Soviet Mercury Grass VHF communications system.

APPENDIX B

TANK AMMUNITION

The ammunition carried by a modern battle tank is highly varied, although its main function is to act against armour; the tank is still considered the best anti-tank weapon. There are two broad types of ammunition: that utilising kinetic energy – where a solid projectile is fired against armour as hard and fast as possible – and that using chemical energy, where an explosive charge is detonated directly against the armour.

The kinetic energy round needs a very high missile velocity to have any effect against modern armour and so a large, long-barrelled high-velocity gun – such as the Chieftain's 120mm L11A5 – is called for. The AP (armour piercing) round itself is made of a very hard material such as tungsten or depleted uranium, capable of smashing its way through hardened steel plate. A number of AP rounds have a small high-explosive charge to increase destructive effect on hitting the target; this round is termed APHE. Despite the hardness of the AP round it tends to shatter on impact. To overcome this problem a penetrative cap is fitted over the nose of the shot (APC – armour piercing capped). This acts as a cushion on impact and yet allows the shot to penetrate the armour. The APC round has a poor ballistic shape, however (it travels badly through the air), and so an APCBC (armour piercing capped, ballistic cap) round is employed which ensures the essential high velocity on hitting the target.

The most important kinetic energy round is the APDS (armour piercing discarding sabot). Within the breech itself a large bore is important so that a powerful charge can propel the round up the barrel at a high velocity; yet when the round hits the target it should have a small diameter in order to increase its penetrative power. The new APFDS (armour piercing fin stabilised discarding sabot) round also ensures that the shot flies true to its target.

Chemical energy rounds have the advantage over kinetic energy ones in that, as high velocity is not important, a smaller, lighter gun or projector can be used. The most simple type is HE (high explosive), which achieves its effect by detonating its charge on the side of the target. Although a very uncomfortable experience for tank crews, HE is not particularly effective against sloped or thick armour; it is mainly used as an anti-personnel weapon, or against soft targets. HEAT (high explosive anti-tank) is more effective, capable of penetrating a considerable thickness of armour and yet needing only a relatively small HE charge to do so. Able to be launched from a simple smooth-bore projector, its main disadvantage is that its 'lethality' is limited: even if a successful hit is made there is a good chance that the tank can survive.

HESH (high explosive squash head) – called HEP (high explosive plastic) by the American armed forces – is a multi-purpose shell, capable of use against a variety of other targets as well as armour. The HESH round must have a low velocity to work and so is less effective when used against fast-moving targets at long range. Another weakness is that, unlike the HEAT round, it is useless when employed against spaced armour. Tanks using the latest Chobham and other similar laminate armours are said to be virtually immune from chemical action AT rounds. Thus the simpler AP rounds will be of increased importance in future tank and anti-tank warfare.

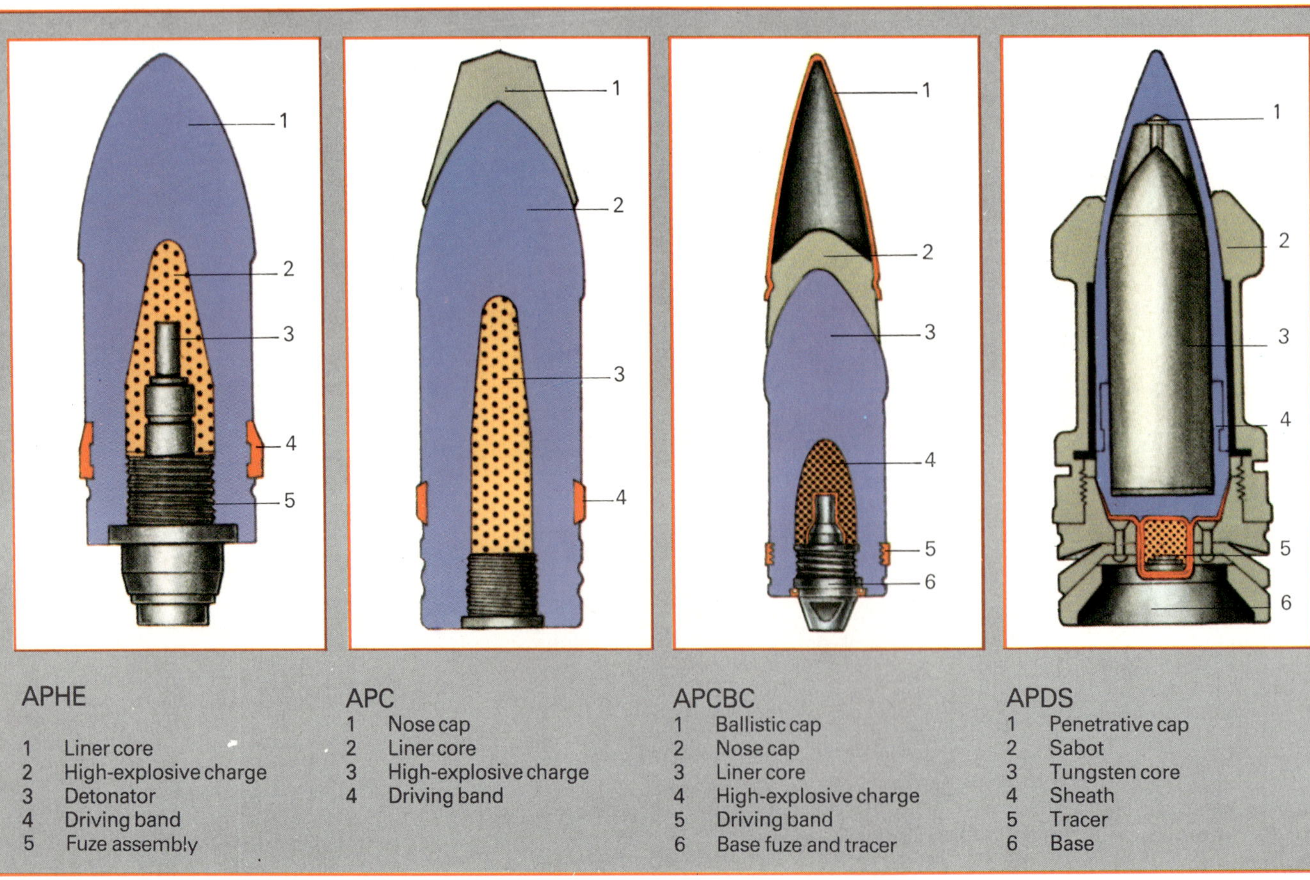

APHE
1 Liner core
2 High-explosive charge
3 Detonator
4 Driving band
5 Fuze assembly

APC
1 Nose cap
2 Liner core
3 High-explosive charge
4 Driving band

APCBC
1 Ballistic cap
2 Nose cap
3 Liner core
4 High-explosive charge
5 Driving band
6 Base fuze and tracer

APDS
1 Penetrative cap
2 Sabot
3 Tungsten core
4 Sheath
5 Tracer
6 Base

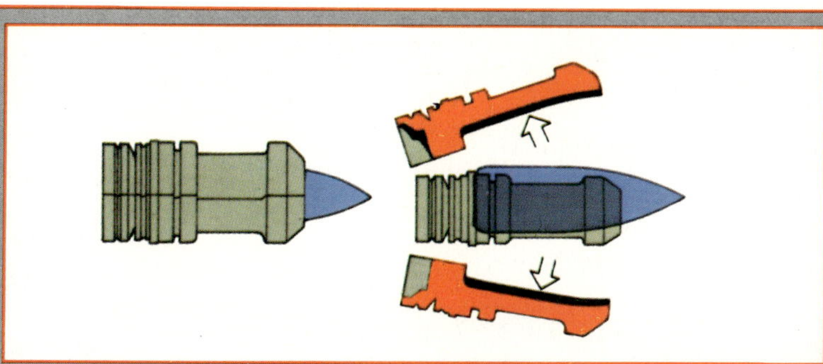

The APDS round consists of a small high-density AP core surrounded by a full-diameter envelope – the sabot. Once the round has left the barrel the sabot disintegrates, leaving the sub-calibre AP core to continue its flight to the target at a very high velocity.

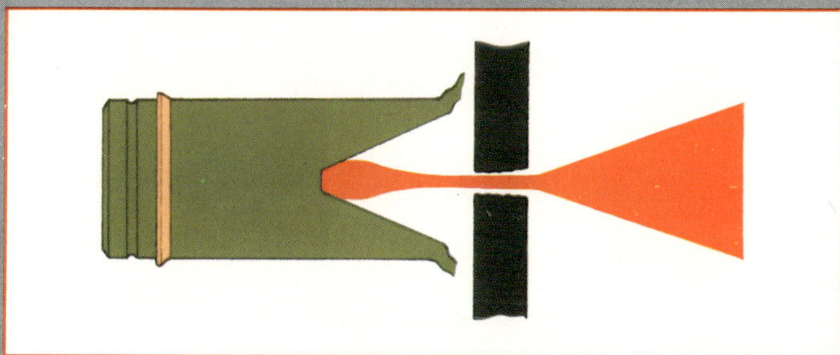

The HEAT round works on the shaped-charge principle: when the round hits its target the exploder charge is detonated, setting off the main charge which is 'funnelled' at extremely high velocity against the armour plate. Achieving a pressure of 200 tonnes per square inch, a jet of molten metal particles cuts its way through the armour.

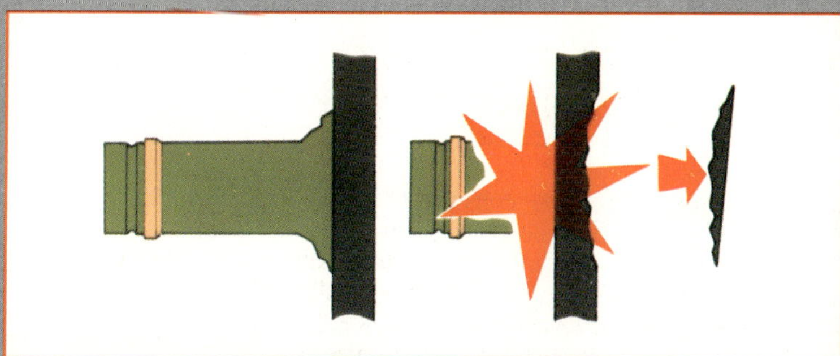

When the HESH round hits the target the explosive charge is 'squashed' against the armour plate, and as the charge is detonated it sends out a series of shock waves which exceed the strength of the plate. A scab of metal on the inside of the armour breaks off at high velocity, causing damage to equipment and personnel within the tank.

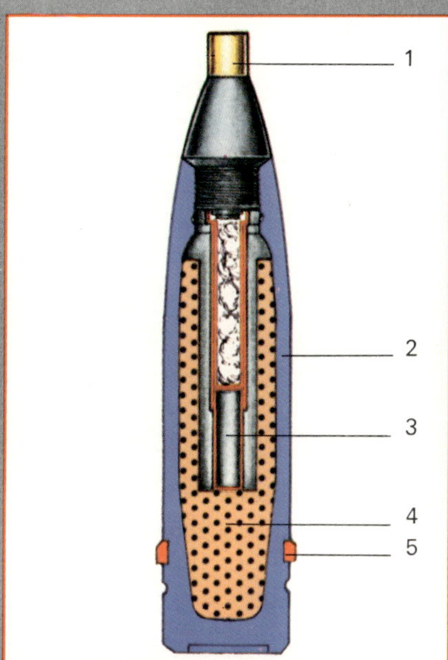

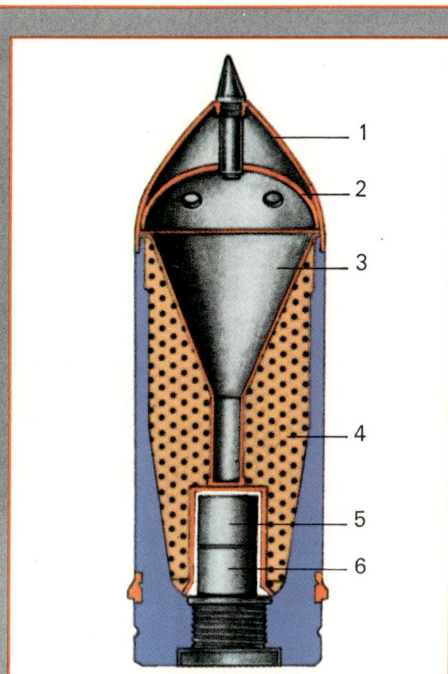

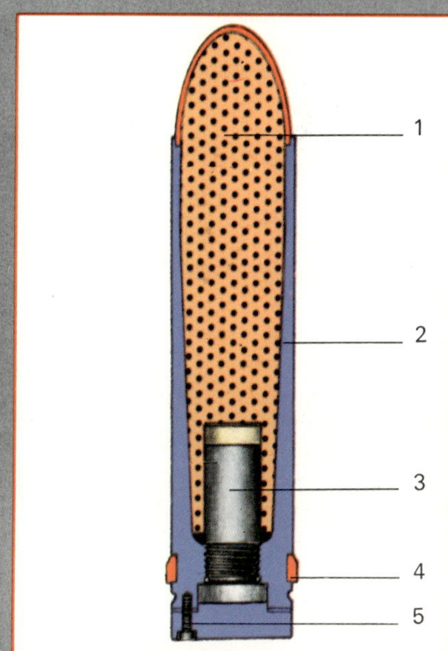

HE
1 Fuze
2 Body
3 Exploder
4 High-explosive filling
5 Driving band

HEAT
1 Nose cap
2 Diaphragm
3 Conical steel liner
4 High-explosive filling
5 Exploder charge
6 Igniter and tracer

HESH
1 High-explosive filling
2 Body
3 Base percussion fuse
4 Driving band
5 Tracer

APPENDIX C

SMALLARMS AMMUNITION

The modern centre-fire cartridge varies widely in size, weight and function, but all types consist of four main components: case, primer, propellant and bullet. The cartridge case, which holds the three other elements, is a cylindrical metal container made either from brass or steel. In the base of the cartridge is the primer which consists of a cap (of a highly sensitive explosive composition) in a recess in the cartridge base. When the firing pin hits the cap a flash is sent forward through one or more holes into the cartridge interior to ignite the propellant. The most common propellant is nitro-cellulose, which comes in flakes, grains or small cylinders, depending on the rate-of-burn required. A powerful combination of nitro-glycerine and nitro-cellulose is sometimes used, but has the disadvantage of causing greater barrel wear. The ignition of the propellant forces the bullet down the barrel at high speed and then out of the gun towards the target.

There is a wide diversity of bullet types depending on the type of weapon and the required function of the bullet. Standard military ammunition for anti-personnel use is ball, which consists of an outer envelope of metal – steel, cupro-nickel or brass – enclosing an inner core of lead alloy or mild steel. Armour-piercing bullets have a special core of hard steel or tungsten carbide. Tracer bullets have a chemical filling in the bullet rear which burns in flight; this is especially useful at night and allows machine gunners to correct their aim while continuing to fire. Observation bullets also use a chemical filling, but fitted in the nose, its function to give a flash or puff of smoke upon impact. This type of bullet is used as a targeting method with larger weapons such as tank guns or recoilless rifles to indicate whether the main gun will hit the target. Another similar bullet is incendiary, which ignites on impact in order to set fire to inflammable objects. High-explosive bullets contain a small explosive charge which is set off by impact.

The designation of cartridge size is derived from the calibre of the bullet, measured as the diameter of the barrel. Calibre designations are not always precise, however; for example, the 7.62mm Nato round has a bullet diameter of 7.8mm. Moreover, although the metric system is standard today, the old imperial designations are still used for many British and US cartridges, for example, the 0.5in (12.7mm) machine-gun round. A fuller description is the calibre plus the cartridge-case length; for example, the US 5.56mm round for the M16 rifle has a case length of 45mm and is consequently known as the 5.56 × 45mm.

The most important smallarms ammunition type is the rifle cartridge which is also used in most machine guns. The traditional high-velocity rifle round was developed in the 1880s and was used virtually unchanged in both world wars. It varied in calibre from 6.5mm to 8mm, could achieve muzzle velocities of between 670 and 760mps (2200–2500fps) and was still accurate at ranges of over 1000m (1100yds).

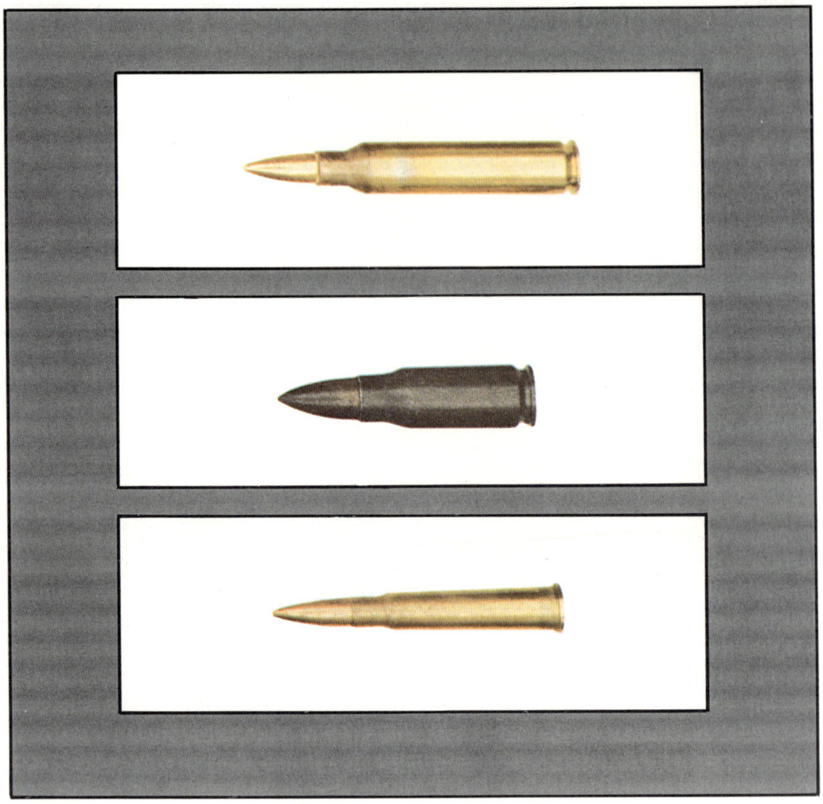

Above: Three assault rifle cartridges. In the centre is the World War II German 7.92 × 33mm round, used in the Stürmgewehr 44. At the bottom is the Soviet 7.62 × 39mm cartridge used in the AK-47, a design based on that of the Stürmgewehr 44. At the top is the first widely-used smaller-calibre round: the US 5.56 × 45mm developed for the M16.

The main weakness of the conventional bolt-action rifle was its relatively low rate-of-fire, and during the inter-war years the major arms-manufacturing nations experimented with self-loading rifles. The Soviet Tokarev and US Garand M1 rifles were the first self-loaders to see widespread service in World War II, and while they increased the rifle section's firepower, they were both heavy, awkward weapons with limited magazine capacities. Because they employed the standard rifle cartridge they had to be large and heavy in order to absorb the bullet's considerable recoil.

The solution lay with a smaller cartridge. Before World War II the German Army realised that as most infantry combat took place at ranges of less than 400m (440yds) the power of the standard cartridge was unnecessary. After experimenting with eight different models they came up with the 'Kurz' (short) cartridge, which was a reduced version of the 7.92 × 57mm 98K round, cut down to 33mm. The Stürmgewehr 44 was then designed around the cartridge, and became the world's first true assault rifle, a lightweight weapon capable of fully automatic fire from a 30-round magazine.

Although the Stürmgewehr 44 saw only limited service during the latter part of the war it made a deep impression on Germany's opponents, who began to copy the rifle or to develop their own designs based on the short cartridge. The Soviet Union introduced the 7.62 × 39mm cartridge towards the end of the war, first used in the Simonov rifle and then in the famous AK-47 assault rifle – the most successful weapon of the postwar period.

The Western Allies, for their part, did not go further than experiment with various cartridge types in the aftermath of World War II. The British set up a research team known as the Small Arms Calibre Panel which came up with a number of innovatory designs. These were not taken up, however: as a

Below: The British 0.303in cartridge is an example of a high-velocity rifle round.

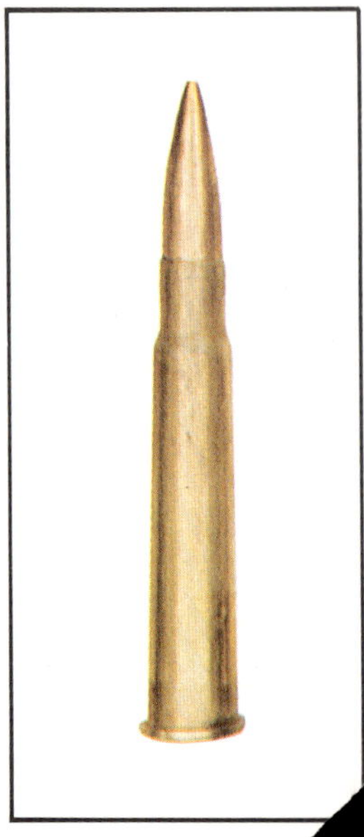

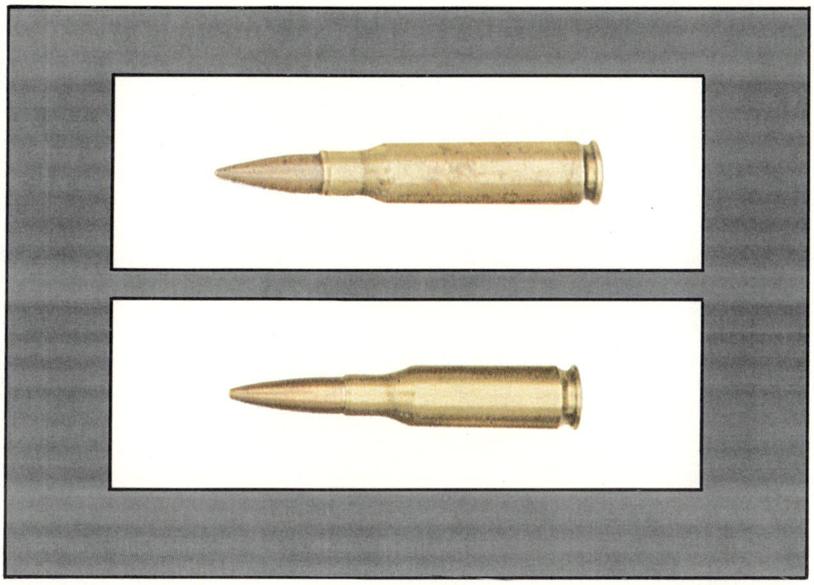

Above: The Nato 7.62 × 51mm round (top) and the British 0.280in EM2 cartridge.

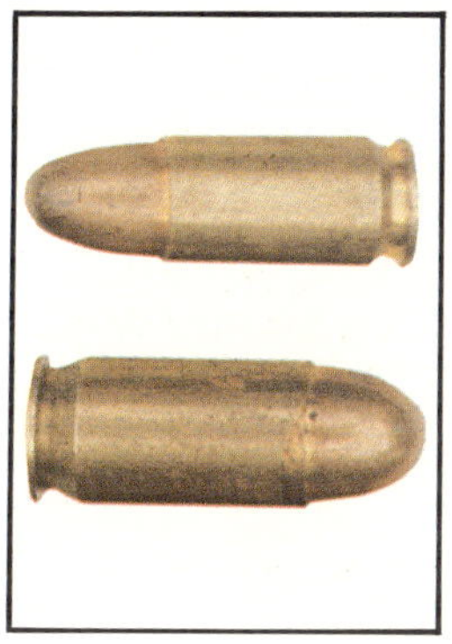

Above: Sub-machine guns use pistol rounds like the German 9mm Parabellum (top) and the US 0.45in round. Below: Heavy machine guns use large cartridges. The Browning M2 fires the 0.5in round (top); the Soviet KPVT the 14.5mm.

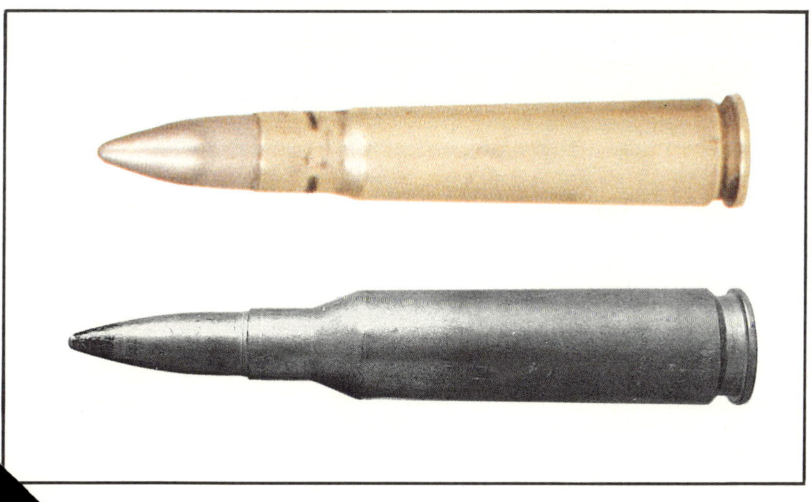

result of the formation of Nato a common cartridge type was deemed essential and the United States had the final say on what would be adopted.

Despite experimentation with short-cased cartridges, a US army requirement that any future round should have the same power (and calibre) as its standard 0.3in cartridge made the lighter cartridge an impossibility. Thus the new Nato round which came into service in 1953 conformed to this stipulation and was fairly conventional, with a case length of 51mm and a calibre of 7.62mm (0.3in). This cartridge became the standard round for the armies of the West, and was used in, among others, the US M14, the Belgian FN series (including the British SLR), and West German G3.

Paradoxically, it was the United States who instigated the next major development in smallarms design, turning its back on the full-calibre, full-length cartridge. The project concluded that a 5.56mm (0.226in) round of sufficient velocity could have as much stopping-power as the 7.62mm round. The 5.56 × 45mm cartridge was consequently developed in conjunction with the AR-15 assault rifle. Initially, the US Army showed little interest in the AR-15, although limited numbers were supplied to the US Air Force and British Army. Eventually the US Army accepted the 5.56mm principle, and the new rifle, designated the M16, was extensively used in Vietnam.

As the United States is the major arms exporter in the West, the M16 soon spread around the world, and several countries began adopting the 5.56mm calibre for their own home-designed weapons. Thus the Israelis manufactured the 5.56mm Galil rifle, the French the MAS assault rifle, the Belgian FN factory brought out a 5.56mm version of the FN, and Heckler and Koch made a similar modification to the G3. The United Kingdom had been an early enthusiast for small-calibre rounds and in the late 1940s had developed the 0.280in EM2. The adoption of the 7.62mm Nato round killed off further British progress, but in the 1970s the new interest in small-calibre cartridges produced the 4.85mm Individual Weapon as a successor to the 7.62mm SLR. The Individual Weapon will equip the British Army from the late 1980s onwards.

By the late 1970s many Nato armies had abandoned the 7.62mm round and so a new series of ammunition trials took place. Although Britain entered the 4.85mm Individual Weapon, the now common 5.56mm calibre was inevitably adopted as the new general standard. The Soviet Union had also taken note of the trend, and brought out the AK-74 with a calibre of 5.45mm.

The development of 5.56mm rounds had been confined to rifles, although a number of light machine guns (LMGs) have now adopted this calibre. On the battlefield of the future the infantry section will be armed with 5.56mm rifles and LMGs while support fire will be provided by 'medium' belt-fed 7.62mm machine guns.

The most recent ammunition development is the caseless round which does away with the conventional cartridge case. Instead of the case the propellant itself is moulded into the right shape to fit into the rifle chamber. The caseless round is at the experimental stage but its advantages in reduced size and weight are considerable.

Two other categories of smallarms ammunition are of note: the pistol and sub-machine gun (SMG) cartridge and the heavy machine gun cartridge. The SMG was first used in limited numbers towards the end of World War I but it only really came into its own in World War II. Produced as a lightweight fully automatic weapon, the SMG was manufactured to supplement the rifle, its function to produce a high volume of firepower at short range. Ammunition was the same as that used in conventional handguns, so that the German MP40 and British Sten used 9mm Parabellum rounds while the Americans used 0.45in pistol ammunition for their Thompson and M3 SMGs. Although the assault rifle has largely replaced the SMG in its old infantry role, large numbers of new models are still made throughout the world.

In the 1980s there was renewed interest in the heavy machine gun. The usual calibre for this type of weapon is 0.5in/12.7mm, although the Soviet Union also uses a 14.5mm cartridge. The main role of the heavy machine gun is as a vehicle-mounted weapon for use against aircraft or thin-skinned armoured vehicles. However, the 0.5in cartridge is highly effective as an anti-personnel round in defensive positions where there is a long field of fire. On a number of occasions during the Falklands Campaign, for example, British troops were brought to a halt by long-range fire from Argentinian 0.5in Browning machine guns.

FURTHER READING

Barker, A. J. *Leopard* (Ian Allan, 1981)
Barnaby, F. (ed.) *Future War* (Michael Joseph, 1984)
Bonds, R. (ed.) *The Chinese War Machine* (Salamander, 1979)
Bonds, R. (ed.) *The Soviet War Machine* (Salamander, 1976)
Campbell, C. *War Facts Now* (Fontana, 1982)
Cockburn, A. *The Threat: Inside the Soviet Military Machine* (Hutchinson, 1983)
Crow, D. (ed.) *Modern Battle Tanks* (Profile Publications, 1978)
Dickson, P. *The Electronic Battlefield* (Boyars, 1976)
Dunstan, S. *Centurion* (Ian Allan, 1980)
Ezell, E. C. *Small Arms of the World* (Stackpole, 1977)
Forty, G. *Chieftain* (Ian Allan, 1979)
Forty, G. *The Scorpion Family* (Ian Allan, 1983)
Foss, C. *Armoured Fighting Vehicles of the World* (Ian Allan, 1982)
Foss, C. *Artillery of the World* (Ian Allan, 1981)
Foss, C. and Gander, T. *Infantry Weapons of the World* (Ian Allan, 1979)
Gander, T. *Encyclopaedia of the Modern British Army* (Patrick Stephens, 1982)
Heitman, H. *South African War Machine* (Galago, 1985)
Hogg, I. *The Encyclopedia of Infantry Weapons of World War II* (Arms and Armour Press, 1977)
Hogg, I. and Weeks, J. *The Illustrated Encyclopedia of Military Vehicles* (Hamlyn, 1980)
Isby, D. C. *Weapons and Tactics of the Soviet Army* (Jane's, 1981)
Lee, R. G. *Introduction to Battlefield Weapons Systems and Technology* (Brassey's, 1981)
Pimlott, J. and Gilbert, A. (eds.) *Modern Fighting Men* (Orbis, 1986)
Suvorov, V. *Inside the Soviet Army* (Hamish Hamilton, 1982)
Tillotson, G. *M48* (Ian Allan, 1982)
Weeks, J. *World War II Small Arms* (Orbis, 1979)